Biological and Abiotic Stress in *Cucurbitaceae* Crops

Edited by Haiping Wang

Published in London, United Kingdom

Biological and Abiotic Stress in *Cucurbitaceae* Crops
http://dx.doi.org/10.5772/intechopen.104275
Edited by Haiping Wang

Contributors
Yalew Yiblet, Agung Wahyono, Titik Budiati, Hafiz Muhammad Shahbaz, Kumud Jarial, Rajender Singh Jarial, Priyanka Bhardwaj, Arpana Sharma, Somya Hallan, Efath Shahnaz, Saba Banday, Ali Anwar, Qadrul Nisa, Gazala Gulzar, Atufa Ashraf, Diksha Banal, Lija Maryomana, Salmabeevi Suhara Beevy, Masoumeh Asadi Aghbolaghi, Mohammad Sedghi, Ghasem Parmoon, Beata Dedicova, Ling-Xi Zhou, Xiang-Dong Li, Chao Geng

First published in London, United Kingdom, 2023 by IntechOpen
IntechOpen is the global imprint of INTECHOPEN LIMITED, registered in England and Wales, registration number: 11086078, 5 Princes Gate Court, London, SW7 2QJ, United Kingdom

British Library Cataloguing-in-Publication Data
A catalogue record for this book is available from the British Library

Additional hard and PDF copies can be obtained from orders@intechopen.com

Biological and Abiotic Stress in *Cucurbitaceae* Crops
Edited by Haiping Wang
p. cm.
Print ISBN 978-1-83769-598-0
Online ISBN 978-1-83769-597-3
eBook (PDF) ISBN 978-1-83769-599-7

Meet the editor

Haiping Wang is a full-time research scientist and professor in the Department of Germplasm Resources, Institute of Vegetables and Flowers, Chinese Academy of Agricultural Sciences (IVFCAAS). He is also the chief of the innovation team of the Department of Germplasm Resources. He conducts research on vegetable genetics to collect germplasm to preserve the diversity of the National Mid-term Genebank for Vegetable Germplasm Resources, China. His areas of interest include genetic diversity and phylogenetic development of vegetable germplasm resources, construction of core vegetable germplasm, and innovative utilization of vegetable germplasm based on modern biotechnology. He also collaborates with the garlic and ginger production industries as well as with consumers. Prof. Wang is the author/co-author of more than 100 publications in scientific journals and 11 book chapters. He is also a reviewer for several journals.

Contents

Preface

The gourd family is one of the most important edible plant families in the world. It includes more than 800 plants, of which the most common species include cucumber, pumpkin, loofah, watermelon, bitter gourd, and so on. *Cucurbitaceae* crops have widespread distribution and important economic value. Some of these crops are cultivated for food or medicine, some have ornamental value, and the fruits have a wide range of applications in daily life.

However, the planting and cultivation of gourd crops are affected by both biological and abiotic stresses, which mainly include various diseases and insect pests. These stresses are becoming more severe year by year and are obstacles to high quality and high yield. Therefore, timely chemical prevention and control should be carried out. Abiotic stress mainly includes extreme temperatures, drought, salinity, nutrient deficiency, heavy metals, and more.

As such, in the planting and cultivation process of gourd plants, it is necessary for growers to pay more attention to disease and pest control, water and fertilizer, light and temperature, and humidity to achieve effective management to improve the yield and quality of gourds. The application of biotechnology to genetic modification of *Cucurbitaceae* plants and the exploration of their own resistance gene sources has broad potential.

This book presents the latest research on biological and abiotic stresses in *Cucurbitaceae* crops. It includes seven chapters organized into four sections. The first section, "Introduction of *Cucurbitaceae* Crops", includes two chapters presenting basic information on *Cucurbitaceae* crops. The second section, "Biological Stress in *Cucurbitaceae* Crops", includes two chapters that discuss various diseases and insect pests that affect *Cucurbitaceae* crops. The third section, "Abiotic Stress in *Cucurbitaceae* Crops", presents some reviews and case studies of abiotic stress in melon and pumpkin. The final section, "Molecular Tools for Studies on *Cucurbitaceae* Crops", discusses the plant virus vectors that could provide molecular tools for studies on Cucurbitaceous plants.

I would like to extend my thanks to all the authors who contributed to this book. My sincere thanks also to Author Service Manager Ms. Zrinka Tomicic at IntechOpen for her assistance throughout the publication process.

Haiping Wang
Institute of Vegetables and Flowers,
Chinese Academy of Agricultural Sciences,
Beijing, China

Section 1

Introduction of *Cucurbitaceae* Crops

Chapter 1

Overview of *Cucurbitaceae* Families

Yalew Yiblet

Abstract

The family *Cucurbitaceae* has a wide range of vegetable or fruit crops that are very important to the national or local economy. Ancient agricultural and medical texts as well as these folktales frequently refer to cultivated members of the *Cucurbitaceae* families. The plants of *Cucurbitaceae* family are rich in phytochemicals such as terpenoids, glycosides, alkaloids, saponins, tannins, and carotenoids responsible for the therapeutic effect. Various parts of these plants exhibit an excess pharmacological activity such as hypolipidemic, anticancer, antidiabetic activity, antimicrobial, anti-inflammatory, and immunomodulatory activities. Among the members of the *Cucurbitaceae* family, pumpkins and squash (*Cucurbita moschata* Duch., *Cucurbita pepo* L., and *Cucurbita maxima* Duch. ex Lam), cucumber (*Cucumis sativus* L.), watermelon (*Citrullus lanatus* L.), and melon (*Cucumis melo* L.) are particularly nutritious due to their beneficial vitamins and minerals. Consumption of some species as food or medicine without proper identification could be dangerous, as some poisonous wild species share a close resemblance with edible ones. A complete, safe, efficient, and cost-effective global conservation system for *Cucurbitaceae* genetic resources should be available, with germplasm and specific accession level information easily accessible, ideally in centralized global databases like Genesys.

Keywords: antimicrobial, antioxidant, *Cucurbitacin*, germplasm, therapeutic

1. Introduction

The *Cucurbitaceae* family is well defined but taxonomically isolated from other plant families. The family contains 115 genera and 960 species that make up this family and are predominantly herbaceous annual vines or perennial lianas and frequently with tendrils [1]. They can be monoecious or dioecious, and occasionally, they are hermaphrodites. They are primarily found in tropical and subtropical zones, and very rarely in temperate zones. The *Cucurbitaceae* are known for their bicollateral vascular bundles, which have phloem on both the outer and inner side of the xylem. *Cucurbitacin*, which is primarily responsible for the bitter flavor, is typically present in *cucurbits* [2]. The family of *Cucurbitaceae* has a wide range of vegetable or fruit crops that are very important to the national or local economy. The vegetables include cucumber (*Cucumis sativus*), zucchini (*Cucurbita pepo*), pumpkin (*Cucurbita maxima*, *C. moschata*, and *Cucurbita argyrosperma*), wax gourd (*Benincasa hispida*), bottle gourd (*Lagenaria siceraria*), bitter gourd (*Momordica charantia*), ridge gourd (*Luffa acutangula*), sponge gourd (*Luffa cylindrica*), chayote (*Sechium edule*), and snake gourd (*Trichosanthes anguina*), and the fruits include melon (*Cucumis melo*), horned cucumber (*Cucumis metuliferus*), watermelon (*Citrullus lanatus*), and luo-hanguo (*Siraitia grosvenorii*). Among these, the bitter gourd (*M. charantia*) and luo-hanguo (*S. grosvenorii*) both have significant culinary and medicinal uses, while snake gourds (*Trichosanthes anguina*) and bottle gourds (*Trichosanthes anguina*) can be utilized both as food and decorative items [3].

2. Origin and history

Plant cultivation started early in northern China. Some domesticates were developed from native species, while others were imported from Southeast Asia. Ancient Chinese myths and stories contain references to numerous vegetable species. Ancient agricultural and medical texts as well as these folktales frequently refer to cultivated members of the *Cucurbitaceae* family. Some of them appear to be indigenous to Indochina, while others were imported from Western Asia and the New World. A recent journey to China and a review of Chinese literature revealed that several cucurbits are consumed in their immature condition as vegetables, some are eaten as fruits, and various species are used for medicinal [4]. In the New World, Africa, the Middle East, the Near East, Asia, and New Guinea, 11 geographical areas have been recognized as the origins of cultivation [5]. Asia is where the cucumis melo originated, and there are many different wild and primitive melons there, especially in India, and wild Melons have been seen in northern Australia, southern America, and northeastern Africa. While archaeobotanists have described the significance of *Cucurbitaceae* plants in prehistoric societies, new molecular data have improved hypotheses about the timing and places of crop domestication as well as revealed the wild ancestors and close relatives of many crops [6].

3. Therapeutic importance of *Cucurbitaceae*

The *Cucurbitaceae* family has several health advantages. Pumpkin has stimulating properties, making it a potential alternative treatment for diseases associated with vertigo [7]. Some nations use the fruits of the *Cucurbita ficifiola* plant, commonly known as leaf gourd, in medicine to treat wounds and hemorrhoids. Phytosterols and tannins are abundant in cucumber. Tannins have astringent qualities, speed up wound healing, and may be able to chelate metal ions [8]. Many phytochemicals in the *Cucurbitaceae* family, including as saponins and cardiac glycosides, frequently used in the treatment of heart problems, have an impact on the circulatory system. Triterpenes, sterol, and alkaloids are among the many bioactive substances they contain high concentrations. Six isoprene units make up the fundamental structure of terpenoids, which are biogenetically descended from active isoprene [9]. The majority of *Cucurbitaceae* family vegetables are high in carotenoids, which are bioactive substances that give them their yellow-red blue. Carotenoids like lutein and zeaxanthin can be found in cucurbits [10]. Saponins have the capacity to coagulate red blood cells, and as a result, they aid in the reduction of bleeding. Due to its ability to suppress lipid oxidation products, primarily malonaldehyde and hydroxynonenal, cucurbitacin also has an anti-atherosclerotic effect. Glycosides are usually used in the treatment of cardiac and heart illness and are present in the leaves and seeds of momoridace balsam. This chemical causes a greater calcium-induced calcium release, which increases the force of heart contraction [11].

3.1 Antioxidant activity

Oxidative stress results from an imbalance between prooxidant and antioxidant levels that favors the former. It is a dangerous state for the entire body and shows signs of human diseases such as cancer or obesity [12]. Cucurbits exhibit antioxidant properties because of a variety of bioactive components, such as cucurbitacins B and E and ellagitannins, which are tannins and have the ability to scavenge free radicals. The majority of vegetables in the *Cucurbitaceae* family are also high in carotenoids, which increase the nutritional value and safety of food due to their antioxidant capacity. A good source of squalene is pumpkin oil. Due to its antioxidant capabilities, this linear triterpene is crucial in preserving the oil's oxidation stability. Bitter gourd seeds, pulp, and its constituent compounds have antioxidant action *via* preventing lipid peroxidation.

3.2 Antidiabetic activity

Diabetes affects more than 380 million individuals in worldwide. There is a significant financial and health load on the health care system. Recent researchers start to look at plants as remedy to treat diabetes including the cucurbit family. Pumpkin has shown that remarkable antidiabetic effects. Pumpkin seed protein-bound polysaccharides had hypoglycemic effects, such as raising plasma insulin levels. Other biologically active substances found in pumpkin include sterols and paraaminobenzoic acid [13]. It also contains a lot of pectin, which helps patients who eat foods high in fiber to control their blood sugar levels and use less insulin. Triterpenoid, steroids, saponins, and alkaloids are only a few of the bioactive phytochemicals that *Momordica charanita* produces. Due to property of polysaccharides in the immature fruits characteristics, the plant is often used in the treatment for diabetes. Even some of the chemicals found in this fruit, such as vicine and charatin, have antidiabetic properties [14].

3.3 Anticancer property

Cancer has several diverse types and one of the most public diseases in the world. Saponins antimutagenic and anticancer properties reduce the risk of cancer. Because they stop a cancer cell's cycle, cucurbitacins produce apoptosis, which has an anticancer effect [15]. Polysaccharides have an antitumor action in addition to acting as hypoglycemic agents. Ergosterol, found in *Mukai maderaspatana*, has anticancer properties. The capacity of *Cucurbita andrena* to inhibit cyclooxygenase demonstrates its anticancer properties. According to [16] who confirmed *C. moschata's* anticancer efficacy against human leukemia cells by acting similarly to ribosome-inactivating proteins, compounds derived from *Curcurbitaceae* family can prevent the formation of cell tumors.

3.4 Anti-inflammatory activities

Cucurbitacins, which are found in cucurbits and are poisonous, have the potential to treat a variety of diseases like inflammation and autoimmune disorders when used in the proper dosage. The most promising *cucurbitacins* B and E have anti-inflammatory activity because they suppress inflammatory mediators including tumor necrosis factor and nitric oxide synthase [17]. In the diet, cucurbits can serve as a source of polysaccharides. These substances have the ability to alter an organism's biological processes because they control macrophages, which in turn affect the immune system and lessen inflammation. *Momoridace balsamina* leaves and seeds contain cardiac glycosides, which protect against fatal endotoxemia and have anti-inflammatory properties. The plant contains ergosterol, a substance with anti-inflammatory and anticancer properties [18].

3.5 Antimicrobial activity

Terpenoids found in abundance in cucurbits damage membranes and limit the growth of bacteria and fungi, which is harmful to them. *Cucumbertacins* is stimulated the stomach secretion. These constituents use an anti-feedant for insects. *Cucurbitacin* have antimicrobial activities that effectively inhibit the growth of *Pseudomonas aeruginosa*, *Escherichia coli*, *Staphylococcus aureus*, and *Bacillus subtilis*. According to [19] the compound isolated from *C. moschata* leaves has an antifungal activity. This significantly reduced *Fusarium oxysporum* growth. As a result of its synergistic interaction with the chitin syntheses inhibitor nikkomycin, isolated protein also reduced the growth of *Candida albicans* [18].

3.6 Hypolipidemic activity

There are numerous reports of plants in the *Cucurbitaceae* family being used therapeutically to treat hyperlipidemia. The lipid-lowering properties of these plants reduce clinical issues associated with lipid metabolism, such as insulin resistance, diabetes, hypertension, and dyslipidemia. By exhibiting lipolysis, decreased lipid content, and decreased mRNA expression of adipocyte transcription factors, *M. charantia* (bitter gourd) fruit juice addresses hyperlipidemia. By controlling the expression of the genes for adipocytokine and adipogenic transcription factors, the accumulation of lipid in primary human adipocytes was reduced.

4. Modern cultivation and production

Citrullus, *Cucumis*, *Cucurbita*, and *Lagenaria* are four of the 115 genera in the *Cucurbitaceae* family that have commercial significance. About half of the world's watermelon (*C. lanatus*) production (29.9 million tons) is produced by four nations: China, Turkey, Iran, and the United States [12]. The watermelon (*C. lanatus*) is an African fruit that was just recently brought to Europe. The second-largest crop grown among the cucurbits is the cucumber (*C. sativus*), and the four nations China, Iran, Turkey, and United States account for 66% of global production. Cucumber was likely domesticated in India and brought to Europe at an early date (during the ancient Greek period). In terms of global production, the top four producers China, Ukraine, Argentina, and Turkey account for 45%. In several different tropical nations, *Lagenaria* has most likely been independently domesticated [12]. In general, productivity is a key factor in determining how much money can be made, but like other vegetable crops, quality, and product availability during times of shortage are also crucial factors in determining how much money can be made. Therefore, efforts should be made to create hybrids and types that are more adaptable during the off-season. Utilizing the known genetic pathways for hybrid seed development would also be relevant to lower the cost of hybrid seeds. High-frequency pistillate lines are encouraged in this regard.

4.1 Germplasm and varieties

Plant viruses are highly reducing crop production globally. There are various ways for crop plants to develop viral resistance. One method is traditional plant breeding, which entails introducing virus-resistance genes from related agricultural plants. Genome editing is a different approach that enables the introduction of genes imparting resistance directly into agricultural plants without the several backcrosses necessary with traditional breeding [20]. As "gene edited crops" do not always contain transgenic portions, they most likely do not require considerable regulation, opening up a new, socially acceptable way to grow crops that are immune to viruses. Among the members of the *Cucurbitaceae* family, pumpkins and squash (*Cucurbita moschata* Duch., *C. pepo* L., *C. maxima* Duch. ex Lam), cucumber (*C. sativus* L.), watermelon (*C. lanatus* L.), and melon (*C. melo* L.) are particularly nutritious due to their beneficial vitamins and minerals. Understanding the information in the genome is crucial for breeders who want to use biotechnology to boost the level of nutrition and the quality of features in *Cucurbitaceae*. Breeding methods that are unconventional can help traditional breeding by reducing costs, prices, and selection effectiveness [21].

4.2 Breeding

Four species of cucurbits the bitter gourd (*M. charantia*), the ridge gourd (*L. acutangula*), the sponge gourd (*L. cylindrica*), and the tropical pumpkin (*C. moschata*) are the focus of current breeding efforts. The choices of species are carefully weighed in a number of variables, such as their economic significance in the targeted geographic areas, nutrient density, and access to genetic resources [22]. The improvement in the crops status on a global scale, production constraints, and the world comparative advantage in resolving them compared to the private sector [23]. For the systematic maintenance and use of the abundance of vegetable crop germplasm, including cucurbits, the Indian Institute of Vegetable Research is a National Active Germplasm Site. Germplasm collection, evaluation, maintenance, and dissemination are the main activities. World Vegetable program stopped cucumber breeding due to the significant private sector investment in this crop, the lack of realistic opportunities for World Vegetable to address production issues, and the low nutritional value of cucumbers in comparison to other vegetable crops. The genetic diversity of current crop species has declined because of domestication and crop improvement bottlenecks. Breeders trying to create superior variations with resilience to both abiotic and biotic stress face a major difficulty due to the limited genetic foundation of many cucurbit crop cultivars. Such genetic diversity may come from crop landraces and wild relatives, which mean they need to be preserved outside of their natural habitat [24]. The comprehensive characterization of gene bank collections is required to speed up the exploitation of conserved diversity in breeding operations. Cucurbits are quite sensitive to a variety of biotic and abiotic stressors. Usually, landraces and their wild cousins contain reservoirs of resistance. Snap melon has been reported to be resistant to fruitfly, downy mildew (*Pseudoperonospora cubensis*), etc. The majority of cucurbit-resistant cultivars were created through straightforward selection. Breeders will be able to create resilient cultivars that are suited to quickly changing environmental conditions by mobilizing a large crop gene pool, which will increase agricultural production and ensure the security of food and nutrition.

5. Poisonous species in the *Cucurbitaceae* family

Without correct identification, it may be unsafe to use some species as food or medication because they may resemble both poisonous and edible wild species. Because incidents of food poisoning in both people and domestic animals have been documented, care must be taken when gathering members of the *Cucurbitaceae* family for food or medicine. In certain instances, species are so similar in vegetative morphology that the flower and fruit characters are the solely means to differentiate between species. Not all of the member of species have had their potential toxicity and safety for human ingestion evaluated. Cucurbitacins are tetracyclic triterpenoids, which are dangerous naturally occurring substances. They are present in several additional *Cucurbitaceae* species as well as *Lufa cylindrica* (L.) M.J. Roem. In the scientific literature, several identified and well-characterized cucurbitacins, including cucurbitacin, have been described. In addition to serving as an insect repellent for the majority of bug species, cucurbitacin is utilized to protect plants from herbivores. The general population needs to be made aware of the dangers of drinking bitter bottle gourd juice and fruit, as the taste is a sign that cucurbitacins are present [25].

African nations such as Botswana, Namibia, Nigeria, Senegal, South Africa, and Swaziland are among those where *C. metuliferus* grows naturally. But in shallow or deep places with well-drained sand, riverbeds, or food plains, this species can thrive. According to certain studies, the unripe fruit of *C. metuliferus* is poisonous, but when completely ripe, it is edible and toxin-free. The fruit's flavor has been compared to a blend of cucumber and banana. When ripe, the fruit is brilliant orange, coated in sharp spikes, and has a bright green, gelatinous flesh. It is frequently used in cooking, as a snack, or eaten raw. In market places in southern African nations such as Zimbabwe, Zambia, Mozambique, Malawi, and South Africa, *C. metuliferus* are frequently found [26].

6. Storage and preservation techniques

A global strategy for protecting *Cucurbitaceae* plant genetic resources should focus on enhancing the effective, economical conservation and utilization of *Cucurbitaceae* plant genetic resources. Ex situ conservation, initiatives deeply on keeping conventional seeds (that can endure long-term freezing or drying) in storage. Although preserving many seeds is an economical conservation technique, it still costs money because they need to be enough infrastructure and staff, especially for the regeneration of old accessions. Optimizing seed viability, seed health, seed storage conditions, and regeneration frequencies is necessary to reduce expenses and the consumption of stored seed. Cucumber, watermelon, and *Cucurbita spp*. seeds, for instance, show less than a 5% reduction in viability in a study of 42 species preserved for 10 years under medium-term conditions (4°C, 30–40% relative humidity, whereas muskmelon showed an 80% decline [26]. The *Cucurbitaceae* strategy survey found that the frequency of such tests varies greatly among gene banks, with the majority-taking place every 5–25 years; however, one gene bank stated that testing took place only extremely infrequently. The ideal testing schedule will vary somewhat depending on the seed storage conditions, which may also affect seed lifetime. A complete, safe, efficient, and cost-effective global conservation system for *Cucurbitaceae* genetic resources should be available, with germplasm and specific accession-level information easily accessible, ideally in centralized global databases like Genesys.

7. Conclusion

The *Cucurbitaceae* family is well known and widely cultivated. Cucumber and pumpkin are the vegetables from this family that are widely consumed, but zucchini and squash are also becoming more common in cooking. Cucurbits have purgative qualities and were used in traditional medicine to cure kidney and bladder stones. In addition to being a good source of vitamins, cucumbers also have medicinal benefits, such as hepatoprotective and anti-inflammatory properties. The *Cucurbitaceae* family, in particular the pumpkin, is known for its antidiabetic properties. In order to cure diabetes, for instance, cucurbits are increasingly being employed for their pharmacological qualities. When it comes to inflammation and low urinary tract disease, some isolated substances, like *M. charantia*, can be *pharmacotherapeutic*. It is essential to do a systematic postharvest analysis of the ideal storage settings and food preparation or processing techniques that will help to maximize the vegetables fruit and lower the levels of cucurbitacin. These actions will assist in maximizing the potential of *Cucurbitaceae* species as a source of local food and medicine. The improvement of the thorough, secure, efficient, and cost-effective conservation and usage of *Cucurbitaceae* plant genetic resources should be the main objective of a global strategy for conserving *Cucurbitaceae* plant genetic resources. Initiatives for ex situ conservation mainly rely on maintaining conventional seeds in storage (seeds that can withstand prolonged freezing or drying).

Conflict of interest

The author declare no conflicts of interest.

Funding

The author received no financial support for this study.

Data availability

The data used to support the finding of this study are included in the chapter.

Author details

Yalew Yiblet
Department of Biology, Natural and Computational Science, Mekdela Amba University, Tulu Awuliya, Ethiopia

*Address all correspondence to: yalewyiblet@gmail.com

References

[1] Paris HS. Historical records, origins, and development of the edible cultivar groups of *Cucurbita pepo* (Cucurbitaceae). Economic Botany. 1989;**43**(4):423-443

[2] Alaziz AA, Attulla FS, Ahmed OK. Effect of winter operating conditions on the performance of a PV/trombe wall: An experimental evaluation. NTU Journal of Renewable Energy. 2022;**2**(1):61-70

[3] Salehi B, Quispe C, Sharifi-Rad J, Giri L, Suyal R, Jugran AK, et al. Antioxidant potential of family Cucurbitaceae with special emphasis on Cucurbita genus: A key to alleviate oxidative stress-mediated disorders. Phytotherapy Research. 2021;**35**(7):3533-3557

[4] Weng Y. Cucumis sativus chromosome evolution, domestication, and genetic diversity: Implications for cucumber breeding. Plant Breeding Reviews. 2021;**44**:79-111

[5] Nerson H. Seed production and germinability of cucurbit crops. Seed Science and Biotechnology. 2007;1 (1):1-0

[6] de Queiroz MA. Germplasm of cucurbitaceae in Brazil. Crop breeding and applied. Biotechnology. 2004;**4**(4):377-283

[7] Ab Azid SM, Wan Rosli WI. Therapeutic benefits of commercially available gourd family in improvement and sustainability of human health. International Journal of Engineering Technology. 2018;**7**(3):164-166

[8] Mukherjee PK, Singha S, Kar A, Chanda J, Banerjee S, Dasgupta B, et al. Therapeutic importance of Cucurbitaceae: A medicinally important family. Journal of Ethnopharmacology. 2022;**282**:114599

[9] Jeffrey C. Notes on Cucurbitaceae, including a proposed new classification of the family. Kew Bulletin. 1962;**15**(3):337-371

[10] Paris HS. Germplasm enhancement of Cucurbita pepo (pumpkin, squash, gourd: Cucurbitaceae): Progress and challenges. Euphytica. 2016;**208**(3):415-438

[11] Ajuru M, Nmom F. A review on the economic uses of species of Cucurbitaceae and their sustainability in Nigeria. American Journal of Plant Biology. 2017;**2**(1):17-24

[12] Olarewaju OO, Fajinmi OO, Arthur GD, Coopoosamy RM, Naidoo KK. Food and medicinal relevance of Cucurbitaceae species in eastern and southern Africa. Bulletin of the National Research Centre. 2021;**45**(1):1-7

[13] Huerta-ReyesM,Tavera-HernándezR, Alvarado-SansinineaJJ,Jiménez-EstradaM. Selected species of the Cucurbitaceae Family used in Mexico for the treatment of diabetes mellitus. Molecules. 2022;**27**(11):3440

[14] Shukla M, Naik VR, Almeida AK, Saklani A. In-vitro screening of cucurbitaceous plants for antidiabetic potential. Indian Journal of Traditional Knowledge. 2019;**18**(2):209-225

[15] Militão GC, Dantas IN, Ferreira PM, Alves AP, Chaves DC, Monte FJ, et al. In vitro and in vivo anticancer properties of cucurbitacin isolated from Cayaponia

racemosa. Pharmaceutical Biology. 2012;**50**(12):1479-1487

[16] Morales-Vela K, Pérez-Sánchez FC, Padrón JM, Márquez-Fernández O. Antiproliferative activity of Cucurbitaceae species extracts from southeast of Mexico

[17] Wahid S, Alqahtani A, Khan RA. Analgesic and anti-inflammatory effects and safety profile of Cucurbita maxima and Cucumis sativus seeds. Saudi Journal of Biological Sciences. 2021;**28**(8):4334-4341

[18] AlJindan R, AlEraky DM, Borgio JF, AbdulAzeez S, Abdalhamid B, Mahmoud N, et al. Diagnostic deficiencies of C. difficile infection among patients in a tertiary hospital in Saudi Arabia: A laboratory-based case series. Saudi Journal of Biological Sciences. 2021;**28**(8):4472-4477

[19] Lira R, Villaseñor JL, Ortíz E. A proposal for the conservation of the family Cucurbitaceae in Mexico. Biodiversity and Conservation. 2002;**11**(10):1699-1720

[20] Attavar A, Tymon L, Perkins-Veazie P, Miles CA. Cucurbitaceae germplasm resistance to verticillium wilt and grafting compatibility with watermelon. HortScience. 2020;**55**(2):141-148

[21] Aruah CB, Uguru MI, Oyiga BC. Variations among some Nigerian Cucurbita landraces

[22] Zampino D, Duro A, Sciandrello S, Parafati L, Restuccia C. Pollen viability and endophytic yeast species of Cistus creticus and C. monspeliensis. Plant biosystems-an international journal dealing with all aspects of. Plant Biology. 2021;**155**(2):384-393

[23] Ma L, Wang Q, Zheng Y, Guo J, Yuan S, Fu A, et al. Cucurbitaceae genome evolution, gene function, and molecular breeding. Horticulture Research. 2022;**9**:12-25

[24] Grumet R, McCreight JD, McGregor C, Weng Y, Mazourek M, Reitsma K, et al. Genetic resources and vulnerabilities of major cucurbit crops. Genes. 2021;**12**(8):1222

[25] Pérez-García A, Mingorance E, Rivera ME, Del Pino D, Romero D, Torés JA, et al. Long-term preservation of Podosphaera fusca using silica gel. Journal of Phytopathology. 2006;**154**(3):190-192

[26] Khan SA, Perveen A. In vitro pollen germination capacity of *Citrullus lanatus* L.,(Cucurbitaceae). Pakistan Journal of Botany. 2010;**42**(2):681-684

Chapter 2

Feasibility of Using Yellow Pumpkin (*Cucurbita moschata*) in Developing Bakery Products

Agung Wahyono, Titik Budiati and Hafiz Muhammad Shahbaz

Abstract

The local abundance of yellow pumpkin (*Cucurbica moschata*) makes it a readily available plant. However, its use as a food item is currently limited to simple processing, resulting in a limited presence of pumpkin-based products in the market. Nevertheless, pumpkin is an agricultural product rich in beneficial components such as high pectin, bioactive substances, beta-carotene, vitamin A, tocopherol, and other vitamins. Its high dietary fiber content, specifically pectin, can help regulate insulin serum levels, lower blood sugar, improve glucose tolerance, and offer protection against various diseases like diabetes, cardiovascular disease, constipation, and colon cancer. In light of the increasing prevalence of conditions like obesity, diabetes, and coronary heart disease, there has been a growing public interest in consuming healthy bakery products. This has led to the development of the bakery industry, with a specific focus on producing healthy and purpose-specific breads. Numerous studies have been conducted to explore incorporating pumpkin into bakery products, as this significantly influences the quality of the resulting bakery products. This chapter will delve into the potential of yellow pumpkin as a nutritious ingredient in the development of bakery products, and its impact on the overall quality of these products.

Keywords: bakery, bread, cake, pumpkin, wheat flour

1. Introduction

Pumpkin is a horticultural crop that belongs to the family *Cucurbitaceae.* Around the world, there are five well-known domesticated species, namely: *Cucurbita moschata* Duchesne ex Poiret, *C. pepo* L., *C. maxima* Duchesne, *C. mixta* pangalo, and *C. micifolia*. Some of them (*C. moschata*, *C. pepo* and *C. maxima*) have great economic value because of their high productivity [1].

Pumpkin is a dicotyledonous seed vegetable that has received great attention because of its nutritional value. The nutritional value of pumpkin is varied to one another depend on the cultivar or species [2]. Generally, pumpkin is cultivated in temperate and subtropical zones of the world. In many countries such as China, India, Yugoslavia, Mexico, America, Argentina, and Brazil, it has been used as a vegetable as well as medicine for therapeutic [1].

IntechOpen

It has been reported that pumpkin is rich in dietary fiber, particularly pectin, functional compounds, bioactive substances, vitamins (A, B_6, K, C, and E) and minerals (K, Mg, P, Se, and Fe) [3]. Some of the bioactive compounds contained in pumpkin are polysaccharides, proteins and peptides, para-aminobenzoic acid, and sterols [4]. Carotenoids are also greatly abundant in pumpkin. Approximately 60 carotenoids have been identified, including b-carotene, a-carotene, and b-cryptoxanthin [5].

In Indonesia, Yellow pumpkin is greatly abundant. But the use of pumpkin is still limited to simple processing. Some processed products of pumpkin have been developed, such as crackers, biscuits, bread, chips, and several types of cakes. However, the availability of processed products of pumpkin in the market is very limited [6]. In Mexico, *Cucurbita ficifolia* is used broadly for several dishes and candies from their fruit or seeds. In Argentina, *C. moschata* is favored by the local community due to its potential for developing salty or sweet food products [1]. In India, immature fruits are cooked as a vegetable, as the same time mature fruit is used to produce confectionery and beverages [7].

Wheat flour is a main ingredient in making bakery products. Until recently, wheat flour was indispensable to making decent quality bakery products. This is due to the ability of wheat flour to form a viscoelastic dough which can retain gas produced by yeast during fermentation. This ability is a key point in producing a good loaf of bread or some bakery products.

The price of wheat flour around the world is affected by supply and demand, wheat productivity, milling requirement, government policies, and economic situation. Considering the unstable market of wheat flour, a growing demand for bakery and confectionary products has resulted in an enthusiastic work to substitute wheat flour with local raw materials.

Blends of other types of flour with wheat flour are known as composite flour. The use of composite flour to produce a leavened or unleavened bakery product has attracted great attention, particularly in countries depending on import to meet the demand of wheat flour [8].

Considering the potencies of pumpkin, a lot of studies have been done in using pumpkin for composite flour in developing bakery products. Ref. [9] reported a study using pumpkin flour to make cookies and muffins and investigated the effect of those products on the hypocholesterolemic, antioxidant, hepatoprotective and prebiotic properties. Ref. [2] developed a bread using a microwave vacuum-dried pumpkin and evaluated its physical, nutritional, and sensorial characteristics. Ref. [3] developed a bread enriched with pumpkin flour and evaluated its physical and structural properties. In addition, another study has been carried out to evaluate its antioxidant and total phenolic contents [6]. Furthermore, a steamed brownies has been developed using pumpkin's premix flour [10].

In general, the use of pumpkin in manufacturing bakery products resulted in decent quality, but some of those parameters are compromised. In this chapter, the feasibility of using pumpkin to manufacture bakery products is reviewed.

2. Bakery

2.1 Definition and history

According to Cambridge's, Merriam Webster's and Collins's dictionaries, a bakery is a place where bread and cakes are made and sometimes sold. Wikipedia describes a bakery or baker's shop more completely as an entity or an establishment that produces and sells flour-based food baked in an oven, such as bread, cakes, pies, cookies, donuts, and pastries.

The history of baking started right after the beginning of recorded history. At that time, our ancestors were able to generate fire from stone. Followed by the discovery of different kinds of grasses that can produce grain for nourishment. From a deliberated experimentation, the dough was created. Baking was done by putting the dough over a heated stone. Nowadays, it is well known as flatbread. Thereafter, the invention of baked products changed the eating habits and lifestyle of our ancestors from being hunters to settlers. Based on archeological evidence shows that baking practices may have been initiated about 23,000 years ago (i.e. ~21,000 BC) during Paleolithic Period [11].

The leavened bread was invented by an accidental work in Egyptian history. The leavened bread was produced with wild yeast from the air and mixed with dough. The dough that had been mixed with wild yeast produced a doubled volume and lighten bread than anyone had ever tasted. For 600 years BCE, the ancient Greeks used an enclosed oven that was heated by fires. People who were willing to bake their bread took the large communal oven. Because the oven was so expensive, they could not afford it personally. Finally, several centuries thereafter, commercial production of bread was performed by ancient Rome, and it became the first baker's profession [12].

2.2 Bakery's products

2.2.1 Bread

Bread is a very popular food around the world regardless of whether those countries produce wheat flour or not. It is a symbol of giving. Bread is prepared in a wide variety form with any meal every day. It is consumed as a snack in many countries as well as a staple food in most countries [12]. According to [13], the term of bread is used to describe a variety of products with different characteristics in shapes, textures, sizes, crusts, colors, softness, eating qualities, and flavors. Basic ingredients of bread are flour and water. Flour is always 100%, and another ingredient is percentage of that of flour weight. In general, bread formula is as follows: 2% of yeast, 4% of sugar, and 2% of sugar and shortening agent [14]. **Figure 1** presented appearances of bread loaves made from various proportions of pumpkin flour.

2.2.2 Cakes

The word of cake came from the Norse word "kaka" of the Viking origin. Cake is like bread but sweeter. For a long time, cake is served during celebration of special event like birthday and weddings. Cake was also closely related to rituals and symbolism from different cultures and countries [12]. For centuries, cake-making has been quite the

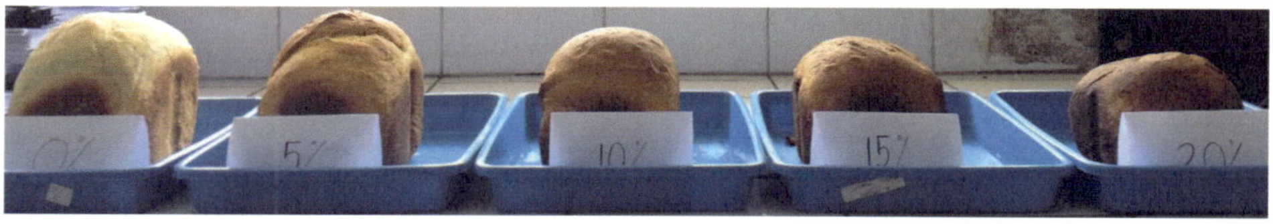

Figure 1.
Appearance of bread loaves made from various proportion of pumpkin flour.

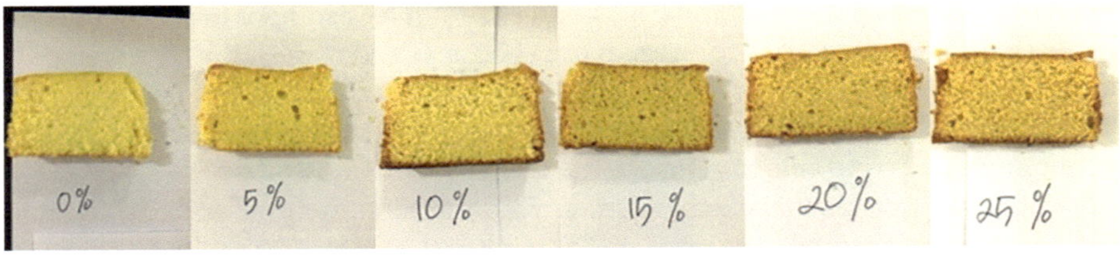

Figure 2.
Cakes made from pumpkin's peel powder.

same. The basic formula of cake comprises of fine wheat flour, refined sugar, butter, and egg. This formula can be modified to produce a wide variety of cakes. Modern cake is defined by a sweet taste, short and tender texture, flavors, and aromas [11]. **Figure 2** presented cakes made from different proportion of pumpkin's peel flour.

2.2.3 Pastries

Pastry made from dough that contained flour, water, and shortening which can be savory or sweetened. If it is sweetened, then we call it baker's confectionery. The terms of pastries are regarding many kinds of baked products made from flour, sugar, milk, butter, shortening, baking powder, and eggs. Pastries are referred to as small tart and other sweet-baked products. Whereas common pastries referred to the dishes include pies, tarts, quiches, and pasties [15, 16]. In addition [11], stipulated that pastries are made by creating alternating layers of dough and fat by folding and rolling the dough.

2.2.4 Biscuits

The name biscuit is synonymous with cookie. Biscuit originated from the word Latin *panis biscotis* meaning twice-cooked bread. This is because the processing of biscuit is divided into two steps by cooking in a hot oven then, followed by drying in cold oven. Biscuit also has two meanings: (i) any of the various small flat sweet cakes, and (ii) small bread leavened with baking powder or soda. In British words, biscuit means a small flattened baked products based on wheat flour with assorted inclusion of fat, sugar, and other ingredients. Today, biscuits are a kind of snack processed with the inclusion of expensive ingredients such as chocolate and cream, resulting luxury gifts, dietary products, also infant food. In general, biscuits are well known as cereal-based products that are baked to a moisture content of less than 5% [11].

2.3 Processing of bakery product

There are some main issues that need to be paid of a lot of attention during the processing of bakery products such as: (i) ingredients should use the exact amount and be measured accurately, (ii) mixing should be done in a correct direction, (iii) pan used in baking should be with regard to the specific product, (iv) correct temperature should be used in baking, (v) pans must be properly prepared for baking to make the baked product easily removed, (vi) preheat the oven to meet the proper temperature for a particular baked product, and (vii) set the temperature and time in accordance to the baked product to get the best quality [17].

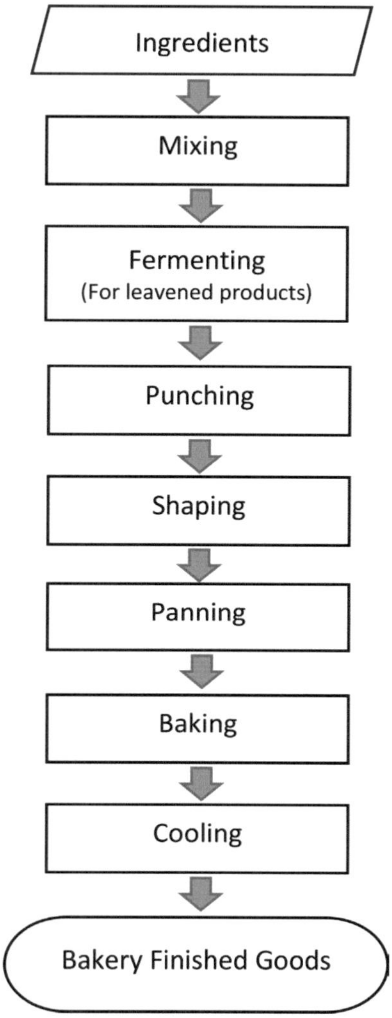

Figure 3.
General processing flows of bakery products.

In general, the processing of bakery products follows several steps, including scaling ingredients, mixing, fermenting (for leavened bread), punching, shaping, panning, baking, and cooling. The general processing flows of bakery's products can be seen in **Figure 3**. Details of the process vary depending on the bakery products. Scaling ingredients must be weighted accurately. Depending on the form of ingredients, scale used to be considered carefully. For example, water, milk, and eggs may be measured by volume. However, if the quantities are large, it is more accurate by weighing. Mixing should be done completely to ensure all ingredients are distributed thoroughly to get a uniform and smooth dough as well as distributing the yeast evenly throughout the dough. Fermentation is the process by which yeast acts on the sugars and starches in the dough to produce carbon dioxide gas and alcohol. It should be aware to not produce an over- or under-fermented dough. An under-fermented dough characterized with un-proper develop volume and coarse texture.

While, over-fermented dough produces sticky, hard-to-work a dough. Punching is hitting the dough by hand to deflate the dough that expels carbon dioxide, redistributes the yeast, relaxes the gluten, and equalizes the temperature throughout the dough. Shaping is performed depending on the bakery products. It can be rounded, flattened, and even un-uniformed. Panning is a process in which the dough is shaped into loaves or rolls and then placed in pans or on baking sheets. Baking is the process in which heat and mass transfer takes place in the dough simultaneously and interdependently and leads to several changes. Cooling is done by removing the baked product from pans and cooled.

3. Quality of bakery products made from pumpkin

3.1 Physical properties

Physical properties are important determinants of food quality. It is related to color, structure, texture, rheology, and interfacial properties. This also applies to bakery products. **Table 1** represents physical properties that are generally evaluated in the bakery products.

In general, study in bakery products examine the physical properties based on the **Table 1**. Physical properties greatly affect the quality of food and can be used to

Product	Physical properties	Methods	References
Bread	Color, measuring luminance (L), red saturation index (a), and yellow saturation index (b)	Color reader	[2]
	Texture, measuring texture profile (TPA)	Texture meter	[18, 19]
	Structure, crumb relative density, and various structural parameters	Image analysis	[20]
	Specific volume, measuring volume-to-weight ratio	Seed displacement methods	[21]
Biscuits	Width and thickness	Scale	[21]
	Spread factor, measuring a ratio between average value of width and average value of thickness	Scale	[21]
Muffin	Specific volume	Seed displacement methods	[21]
	Water absorption index (WAI) and water solubility index (WSI)	Gravimetry	[22]
	Color	Color reader	[22]

Product	Physical properties	Methods	References
Cake	Volume	Seed displacement methods	[23]
	Texture	Texture analyzer	[23]
	Color	Color reader	[24]
	Structure	Image analysis	[24]
Baked rolls	Volume and specific volume	Seed displacement methods	[21]
	Texture	Texture analyzer	[25]

Table 1.
Physical properties of bakery products and methods for determination.

classify and identify them. In a globalized market, the differentiation of food must be based on its physical properties [26]. There have been many studies that have reported the use of pumpkins in developing bakery products. Mostly, incorporation of pumpkin significantly affects the physical properties of bakery products. In baked roll, increased proportion of pumpkin powder resulted in a lower volume and specific volume. The texture of the enriched baked roll also firmer compared to that of the control during storage [25]. In bread, increasing the dried pumpkin proportion resulted in decreasing bread quality. It became unacceptable because of worse porosity and stickier bread crumb. The intensity of the yellow color (*b* value) was more pronounced than that of the control bread. The intents yellow color of bread was contributed from pumpkin color [2]. Ref. [3] reported that at 10% and higher levels of pumpkin lead to decreasing specific volume of enriched bread. Pumpkin enrichment did not affect the texture of bread compared to that of the control bread. The crumb luminance decreases significantly with 10% of pumpkin flour. The addition of pumpkin flour was also significantly affected the crumb structure of bread. Mean cell area of enriched breads became smaller than those of control breads. The crumb feels to the mouth are greatly associated with the cell structure of the crumb. The finer, thin-walled, uniform cells size produced a softer and more elastic texture than that of the coarse and thick-walled cell structures. Ref. [27] reported that progressive addition of pumpkin flour giving bread with relatively low specific volume. Progressive addition resulted in an initial rise and subsequent decrease in loaf volume. Suggested addition of pumpkin powder is up to 10%. In muffins, the partial replacement of wheat flour with pumpkin flour had no significant effect on the firmness and image analysis such as distribution, size, and pore area of the crumb. However, the partial replacement resulted in a darker crust color, higher yellowness, and a lower specific volume compared to that of control muffins. In biscuits, there are three parameters related to the physical properties of biscuit, namely width, thickness, and spread factor. Thickness is a quality parameter of biscuit, which represents height of biscuit. Its optimum value is always desirable by manufacturer. The spread factor is a ratio between the average value of the width and average value of thickness. The addition of pumpkin flour resulted in a significant decrease in width and spread factor. On the other hand, thickness was significantly increased. The decrease in width values of biscuits might be contributed by a lesser water absorption of the blended flour. The increase in thickness of biscuits made from blended flour might be due to the dilution of gluten contents.

The leverage of water absorption in blended flour might produce a greater viscosity of dough which can trigger the increase in thickness of the biscuits reducing its volume spread [28].

3.2 Chemical properties

The chemical properties of food are generally related to the nutritional values of food. In many cases, chemical properties are more important compared to those of physical properties. Some vegetables are indispensable in producing a wide range of bread products because of their chemical compositions [29]. The chemical properties include macronutrients such as saccharides, proteins, fats, and micronutrients such as minerals, vitamins, colorants, additives, fibers, and phytochemicals. In general, analysis of chemicals properties of bakery products is listed in **Table 2**.

Product	Chemical properties	Methods	References
Bread	Total phenolic contents	Spectrophotometry using standard curve	[6]
	Antioxidant	DPPH and ABTS scavenging	[18]
	Protein	Kjeldahl method	[29]
	Moisture content	Oven-drying methods	[2]
	Vitamin C contents	Iodometric	[2]
	Vitamin contents	HPLC	[29]
	Carotenoid	Spectrophotometric	[2]
	Reducing sugar	LVS 252:2000	[2]
	Total fat	ISO 6492:1999	[2]
	Total fat	Soxhlet	[29]
	Minerals	Atomic absorbance spectrometer	[29]
	Ash	Muffle furnace (gravimetry)	[29]
Muffin	Moisture content	Oven-drying	[30]
	Ash	Dry-ashing	[30]
	Fat	Soxhlet	[30]
	Protein	Kjeldahl	[30]
	Carbohydrate	By difference	[31]
	Total dietary fiber	Enzymatic	[31]
Biscuit	Moisture content	Oven-drying	
	Ash	Dry-ashing	[32]
	Insoluble, soluble, and total dietary fiber	Enzymatic	[32]
	Protein	Semi-micro Kjeldahl	[32]
	Fat	Soxleth	[32]
Cake	Moisture contents	Oven-drying	[33]
	Β-carotene	Spectrophotometric	[34]

Product	Chemical properties	Methods	References
	Ash	Dry-ashing	[34]
	Crude fat	Soxleth	[33]
	Crude fiber	Enzymatic	[33]
	Carbohydrate	By difference	[33]
	Protein	Micro-Kjeldahl	[33]

Table 2.
Chemical properties of bakery products and methods for determination.

In bread, carotenoid contents increased 5.5 times higher with the addition of pumpkin flour compared to that in control bread. This is due to an abundant proportion of carotenoids in pumpkins. There was no significant difference in moisture content between control bread and bread with pumpkin flour. This is because the moisture content of bread with and without pumpkin addition mainly depends on the amount of water during dough making. The reduced sugar content of bread with pumpkin addition was 1.6 times higher compared to that in control bread. This result explained that higher reduced sugar content in pumpkin bread contributed to a high reduced sugar content in pumpkin flour. Fat content and vitamin C content were comparable between pumpkin bread and control bread [2]. Pumpkin flour significantly enhanced the antioxidant activity of enriched bread. The highest antioxidant activity was observed in bread with an enrichment of 20% of pumpkin flour. The addition of pumpkin flour produced higher levels of β-carotene. Pumpkin flour contained about 180 μg/100 g β-carotene. It is higher than that of wheat flour which does not contain vitamin A. β-Carotene has the ability to be an antioxidant that can play an important role in stabilizing carbon nucleated radicals. β-Carotene in pumpkin flour acted as a provitamin and exhibited an antioxidant activity. Antioxidant activity is also influenced by several components, including phenol compounds which are the basic framework of compounds that have antioxidant activity. Increased levels of pumpkin flour produced a higher level of total phenolic content in enriched bread. The total phenolic content of pumpkin flour is affected by the drying of pumpkin chips during processing. The high level of temperature used during drying leads to the formation of phenolic compounds. The total phenol is directly proportional to the antioxidant activity [6].

In muffins, the addition of 5 g/100 g pumpkin flour produced a comparable level of protein, digestible carbohydrates, and lipids compared to those in the control muffin. However, the enriched muffins showed a higher content of ash and dietary fiber [22].

The incorporation of pumpkin flour for the preparation of the cake produced higher moisture, crude fiber, ash, and β-carotene contents of the resulting cake. On the contrary, the crude protein, crude fat, and carbohydrate in cake decreased. The increase in moisture content might be due to the hygroscopic nature of pumpkin powder and wheat flour and the higher water absorption capacity in the composite flour compared to wheat flour. The higher crude fiber in pumpkin cake was contributed from the highly insoluble dietary fiber which includes cellulose, hemicellulose, and lignin in pumpkin flour. The higher ash and β-carotene of pumpkin cake was attributed to the high ash and β-carotene content in pumpkin flour [34]. In biscuits, the nutritional quality was positively influenced by the incorporation of pumpkin. The incorporation of pumpkin increased the protein, crude fiber, calcium, carotene, and vitamin C of biscuits. This was because those components were found to be in higher amounts in pumpkin puree. In addition, wheat flour is considered not a good source of carotene and vitamin C [35].

3.3 Sensory properties

Sensory evaluation is a scientific method performed to evoke, measure, analyze, and interpret those responses to products as perceived through the senses of smell, touch, taste, sight, and hearing [36]. Healthy food has attracted a great attention for years. Therefore, many studies have been performed to use healthy ingredients like pumpkin to develop new products. Consumers want their food to be healthy, but when it comes down to making a buying choice, it is taste that matters most [37]. In bakery-enriched pumpkin products, the physical and chemical properties are not the only contributor for bakery quality. The sensory properties are also important to contribute to overall bakery quality. Usually, the sensory properties are studied through consumer research. Consumer research is important to know the general acceptance of consumers to the newly developed product. By this activity, companies or factories determine consumer liking, preference, and opinions on the newly developed products. Finally, this information can be used to decide such as the production and marketing of new products, the reformulation of existing product, the acceptance of suppliers and processes, and the establishment of quality control specifications [2].

In bread, consumer rated higher on bread with a pumpkin than that of control bread. The rated score was 7.3 for bread with pumpkin and 6.7 for the control bread, respectively. A bread sample with pumpkin is tastier than a control bread sample [2]. It has been reported that the increases in loaf volume of bread with pumpkin were accompanied by substantial increases in organoleptic acceptability. It showed an essentially linear dependence on the specific volume up to 4.3 ml/g [27].

The use of pumpkin in muffin rated a higher acceptance by consumers. A 5% of pumpkin addition was preferred by consumers than that of 10% of addition in muffin. The 10% addition of pumpkin powder resulted in a pronounced or strong pumpkin flavor, a strong yellowish color, and a dry texture [22].

In cookies, 10% of pumpkin powder had no significant difference in total acceptance from that of control cookies. Increased levels of pumpkin powder concurrently decreased the consumer acceptance of enriched cookies. In chiffon cake, 20% of pumpkin powder resulted in a comparable acceptance to that of control cake. A higher level of pumpkin powder produced a cake with a lower acceptance. Interestingly, in butter cake, up to 50% of pumpkin powder enrichment had no effect on consumer acceptance compared to that of control cake [38].

4. Conclusions

The utilization of pumpkin in various forms significantly impacts the physical, chemical, and sensory characteristics of bakery products. Typically, when pumpkin is incorporated into bakery items, it leads to a decrease in specific volume, a firmer texture, and a darker appearance. However, the enrichment of bakery products with pumpkin enhances their nutritional value by adding vitamin C, β-carotene, phenolic compounds, minerals, and dietary fiber. Furthermore, the sensory attributes of bakery products are strongly influenced by the inclusion of pumpkin, with the extent of impact depending on the proportion of pumpkin used in the recipe. In general, bakery products remain acceptable when the pumpkin content does not exceed 20%. Consequently, it can be concluded that incorporating pumpkin into bakery products is a feasible option, albeit with certain compromises in terms of their physical and sensory properties.

Acknowledgements

The authors would like to thank the Ministry of Education, Culture, Research and Technology, the Republic of Indonesia, and Kyungpook National University, the Republic of South Korea.

Author details

Agung Wahyono[1*], Titik Budiati[1] and Hafiz Muhammad Shahbaz[2]

1 State Polytechnic of Jember, Jember, Indonesia

2 University of Veterinary and Animal Science, Lahore, Pakistan

*Address all correspondence to: agung_wahyono@polije.ac.id

References

[1] Zhou T, Kong Q, Huang J, Dai R, Li Q. Characterization of nutritional components and utilization of pumpkin. Food. 2007;**1**:313-321

[2] Rakcejeva T, Galoburda R, Cude L, Strautniece E. Use of dried pumpkins in wheat bread production. Procedia Food Science. 2011;**1**:441-447. DOI: 10.1016/j.profoo.2011.09.068

[3] Wahyono A, Tifania AZ, Kasutjianingati KE, Kang WW, Chung SK. Physical properties and cellular structure of bread enriched with pumpkin flour. In: Proceedings of IOP Conference Series: Earth and Environmental Science. 2018;**207**(1):1-6. DOI: 10.1088/1755-1315/207/1/012054

[4] Adams GG, Imran S, Wang S, Mohammad A, Kok S, Gray DA, et al. The hypoglycaemic effect of pumpkins as anti-diabetic and functional medicines. Food Research International. 2011;**44**(4):862-867. DOI: 10.1016/j.foodres.2011.03.016

[5] Murkovic M, Mülleder U, Neunteufl H. Carotenoid content in different varieties of pumpkins. Journal of Food Composition and Analysis. 2002;**15**:633-638. DOI: 10.1006/jfca.2002.1052

[6] Wahyono A, Dewi AC, Yudiastuti SON, Jamilah S, Kang WW. Antioxidant activity and total phenolic contents of bread enriched with pumpkin flour. IOP Conference Series: Earth and Environmental Science. 2020;**411**(1): 1-7. DOI: 10.1088/1755-1315/411/1/012049

[7] Mala SK, Kurian AE. Nutritional composition and antioxidant activity of pumpkin wastes. International Journal of Pharmaceutical, Chemical and Biological Science. 2016;**6**(3):336-344

[8] Aziah AAN, Komathi CA. Physicochemical and functional properties of peeled and unpeeled pumpkin flour. Food. 2009;**74**(7):328-333. DOI: 10.1111/j.1750-3841.2009.01298.x

[9] Dyshlyuk L, Babich O, Prosekov A, Ivanova S, Pavsky V, Yang Y. In vivo study of medical and biological properties of functional bakery products with the addition of pumpkin flour. Bioactive Carbohydrates and Dietary Fiber. 2017;**12**:20-24. DOI: 10.1016/j.bcdf.2017.09.001

[10] Subaktilah Y, Wahyono A, Yudiastuti SON, Avianti TM. Chemical characteristic of steamed pumpkin brownies premix flour. In: Proceeding of IOP Conference Series: Earth and Environmental Science. 2021;**672**(1): 1-4. DOI: 10.1088/1755-1315/672/1/012054

[11] Zhou W, Therdthai N, Hui YH. Introduction to baking and bakery products. In: Zhou W, Hui YH, De Leyn I, Pagani MA, Rosell CM, Selman JD, et al, editors. Bakery Products Science and Technology. 2nd ed. New York: Wiley; 2014. p. 4-13. DOI: 10.1002/9781118792001

[12] Uttarakhand Open University. Bakery and Confectionary. Available from: https://uou.ac.in/sites/default/files/slm/HM-302.pdf [Accessed: March 17, 2023]

[13] Cauvain SP. Bread making: An overview. In: Cauvain SP, editor. Bread Making. Woodhead Publishing Limited, Boca Raton: CRC; 2005. pp. 7-26

[14] Mondal A, Datta AK. Bread baking. A review. Journal of Food Engineering. 2008;**86**:465-474

[15] Friberg B. The Professional Pastry Chef: Fundamentals of Baking and Pastry. 4th ed. New York: Wiley; 2002. p. 1040

[16] Coyle LP. The World Encyclopedia of Food. 1st ed. New York: Facts On File Inc; 1982

[17] Patil H. Introduction to Bakery & Confectionery. 2020. Available from: https://aissmschmct.in/wp-content/uploads/2020/07/Chapter1-Introduction-to-bakery-confectionery.pdf [Accessed: Febraury 2, 2023]

[18] Wahyono A, Kang WW, Park HD. Characterization and application of *Torulaspora Delbrueckii* JK08 and *Pichia Anomala* JK04 as Baker's yeast. Journal of Food and Nutrition Research. 2015;**54**:205-217

[19] Ulziijargal E, Yang JH, Lin LY, Chen CP, Mau JL. Quality of bread supplemented with mushroom mycelia. Food Chemistry. 2013;**138**:70-76

[20] Zghal MC, Canlon MG, Sapirstein HD. Cellular structure of bread crumb and its influence on mechanical properties. Journal of Cereal Science. 2002;**36**(2):167-176. DOI: 10.1006/jcrs.2001.0445

[21] AACC. Approved Methods of the AACC. 10th ed. St. Paul: American Association of Cereal Chemists; 2000

[22] Scarton M, Nascimento GC, Felisberto MHF, Moro TMA, Behrens JH, Barbin DF, et al. Muffin with pumpkin flour: Technological, sensory and nutritional quality. Brazilian Journal of Food Technology. 2021;**24**:1-12. DOI: 10.1590/1981-6723.22920

[23] Ben Jeddou K, Bouaziz F, Zouari-Ellouzi S, Chaari F, Ellouz-Chaabouni S, Ellouz-Ghorbel R, et al. Improvement of texture and sensory properties of cakes by addition of potato Peel powder with high level of dietary Fiber and protein. Food Chemistry. 2017;**217**:668-677. DOI: 10.1016/j.foodchem.2016.08.081

[24] Rios RV, Garzón R, Lannes SCS, Rosell CM. Use of succinyl chitosan as fat replacer on cake formulations. LWT - Food Science and Technology. 2018;**96**:260-265. DOI: 10.1016/j.lwt.2018.05.041

[25] Lucia M, Lauková M, Karovičová J, Kohajdová Z. Utilization of pumpkin powder in baked rolls. Potravinarstvo Slovak Journal of Food Sciences. 2018;**12**(1):195-201. DOI: 10.5219/887

[26] Arana I. Basics of electronic, nondestructive technologies for the assessment of physical characteristics of foods. In: Arana I, Editor. Physical Properties of Food. Boca Raton. CRC Press; 2012. p. 1-22

[27] Ptitchkina NM, Novokreschonova LV, Piskunova GV, Morris ER. Large enhancements in loaf volume and organoleptic acceptability of wheat bread by small additions of pumpkin powder: Possible role of acetylated pectin in stabilising gas-cell structure. Food Hydrocolloids. 1998;**12**(3):333-337. DOI: 10.1016/S0268-005X(98)00024-1

[28] Hussain A, Kausar T, Aslam J, Quddoos MY, Ali A, Kauser S, et al. Physical and rheological studies of biscuits developed with different replacement levels of pumpkin (*Cucurbita Maxima*) Peel, flesh, and seed powders. Journal of Food Quality. 2023;**2023**:1-13. DOI: 10.1155/2023/4362094

[29] Bayramov E, Aliyev S, Gasimova A, Gurbanova S, Kazimova I. Increasing the biological value of bread through the application of pumpkin puree.

Eastern-European Journal of Enterprise Technologies. 2022;**2**(11-116):58-68. DOI: 10.15587/1729-4061.2022.254090

[30] AOAC. Official Methods of Analysis of Association of Official Analytical Chemist. Virginia USA: AOAC International; 2005

[31] Rismaya R, Syamsir E, Nurtama B, Tohyeng N. Effects of water addition and baking time on the production process optimization of pumpkin muffins: Pilot plant study. Canrea Journal: Food Technology, Nutritions, and Culinary Journal. 2022;**5**(2):183-207. DOI: 10.20956/canrea.v5i2.711

[32] AOAC. Official Methods of Analysis of Association of Official Analytical Chemist. Virginia USA: AOAC International; 1996

[33] AOAC. Official Methods of Analysis of Association of Official Analytical Chemist. Virginia USA: AOAC International; 1995

[34] Bhat MA, Bhat A. Study on physico-chemical characteristics of pumpkin blended cake. Journal of Food Processing & Technology. 2013;**4**(9). DOI: 10.4172/2157-7110.1000262

[35] Gurung B, Ojha P, Subba D. Effect of mixing pumpkin puree with wheat flour on physical, nutritional and sensory characteristics of biscuit. Journal of Food Science and Technology Nepal. 2016;**9**:85-89

[36] Kemp SE, Hollowod T, Hort J. Sensory Evaluation: A Practical Handbook. New York: Willey-Blackwell; 2009. p. 208

[37] Simons C. Sensory Properties of Food. Available from: https://cwsimons.com/sensory-properties-of-food [Accessed: April 27, 2023]

[38] Pongjanta J, Naulbunrang A, Kawngdang S, Manon T, Thepjaikat T. Utilization of pumpkin powder in bakery products. Songklanakarin Journal of Science and Technology. 2006;**28**(1):71-79

Section 2

Biological Stress in *Cucurbitaceae* Crops

Chapter 3

Bacterial Leaf Spot of Cucurbits: A Menace to Cultivation

Kumud Jarial, Rajender Singh Jarial, Priyanka Bhardwaj, Arpana Sharma and Somya Hallan

Abstract

Bacterial leaf spot is one of the important diseases of cucurbits affecting almost all the cultivated cucurbits worldwide. The disease has been reported to cause huge losses to the cucurbits as the symptoms of the disease appear on all plant parts including fruits. The disease is favoured by moderate-to-high temperatures and high humidity. The pathogen perpetuates in the seed and infected crop debris. The management efforts for the disease are still on and there are reports of its management up to more than 50 per cent level under field conditions. No resistant variety to this disease is known. However, the disease can be managed by the use of common antibiotics. The disease is an emerging disease at international and national level. Keeping in view the importance of the disease, the chapter is discussed under following headings such as occurrence, losses, symptoms, pathogenicity and host range, cultural studies, molecular characterisation, survival, disease cycle and epidemiology, host resistance and management, covering all the important aspects in the light of available literature.

Keywords: Xanthomonas cucurbitae, bacterial leaf spot, cucurbits, symptomatology, management

1. Introduction

Family *Cucurbitaceae*, the second largest family among fruits and vegetables next to Solanaceae, contains genetically a diverse group of plants, and several important commodity crops in many parts of the world such as cucumber, pumpkin and melon [1]. Cucurbits, also known as gourds, consist of hundred genera, most of which are edible worldwide [2]. The most important edible plants of the family are Cucumber, pumpkin, different types of gourds, melons, squash and zucchini. The Cucurbit family is a source of dietary fibre such as β-carotene pro-vitamin A, potassium and vitamin C [3]. It is also a rich source of phytochemicals such as cucurbitacins, saponins, carotenoids, phytosterols, polyphenols and antioxidants [4]. Fruit can be consumed ripe (e.g. pumpkin) or unripe (e.g. zucchini), raw (watermelon), cooked (squash) or pickled (gherkins) and is a type of berry known as pepo [5]. Also, cucurbits are consumed fresh for the purpose of dessert (muskmelon and watermelon), in salad (cucumber and long melon), cooked (bottle gourd, bitter gourd, pumpkin, etc.), jam (pumpkin) or candied (ash gourd). Pumpkin which is also known by other names

 IntechOpen

such as 'Sitaphal', 'Kashiphal' or 'kaddu' [6] is used in traditional medicine systems like antidiabetic, antihypertensive, antitumor, antibacterial, antihypercholesterolemia, intestinal antiparasitic and anti-inflammation [7]. Bottle gourd is one of the largest produced cucurbit vegetables and is also called calabash gourd, trumpet gourd, white-flowered gourd and zucca melon [8]. Bitter gourd has high nutritional value due to the presence of ascorbic acid and iron contents [9]. Watermelon flesh is thirst-quenching, highly nutritious and has vitamins C and A in the form of beta-carotene [10]. Luffa (Luffa cylindrica (L.)) is commonly called sponge gourd, loofah, vegetable sponge, bath sponge or dishcloth gourd [11]. Cucumber (Cucumis sativus L.) is the fourth most important vegetable worldwide [12].

Cucurbits are attacked by number of insects and pathogens. Among these, Xanthomonas cucurbitae (Bryan) Dowson or Xanthomonas campestris pv. *cucurbitae* causing bacterial spot is emerging as an important pathogen, which is a menace to cucurbits cultivation, especially in case of pumpkin, winter squash and bottle gourd and causes huge losses in the crops of up to 90 per cent [13]. About 100 per cent loss in yield of pumpkin is also reported [14]. The disease is Nationally and internationally important especially in bottle gourd and pumpkin, respectively [15]. This disease is also reported as an emerging disease of cucurbits [16].

2. Occurrence

Bacterial leaf spot disease was reported for the first time on Hubbard squash in New York in 1926 [17]. The disease has since then been reported in various cucurbits [18] like on watermelon from Georgia [19]; pumpkin from the United States [20], Reunion Island [21], Nepal [22], and Italy [23] and bottle gourd from India [13]. In India, the disease was first reported in 1989 from Bihar on cucumber. After that, there has no report of the disease in India till 2011 [13]. A survey was conducted in 64 areas including five regions of Himachal Pradesh, India, to reveal disease severity in the range of 12.50 to 78.33 per cent in bottle gourd [13]. A similar survey was conducted in Illinois, USA, between 2010 and 2011 in pumpkin fields, which demonstrated the prevalence of bacterial spots in the field on infected fruits with 25 per cent disease incidence on an average. The survey in Jammu and Kashmir, India, including 27 areas revealed disease severity ranging from 14.5 to 33.4 per cent [24]. Bacterial leaf spot was also recorded to be a predominant disease in 60 per cent of fields of pumpkin surveyed on Caribbean Island with 30 per cent disease incidence in the rainy season [25].

3. Losses

Xanthomonas cucurbitae has been reported to cause significant losses in cucurbits throughout the world. X. cucurbitae is prevalent during warm and humid conditions on different cucurbits. This pathogen can reduce the yield of the crop by up to 90 per cent by causing severe infection of foliage as well as fruit. Yield losses of up to 20 per cent have been observed in highly susceptible cultivars with the disease severity sometimes reaching up to 50–60 per cent during storage [26]. Ninety per cent yield loss has been reported in pumpkin fields in Illinois. [27] Bacterial leaf spot caused by X. cucurbitae is an important disease of cucurbits leading to huge crop losses especially to bottle gourd, pumpkin and squashes in sub-tropical zone of Himachal

Pradesh, India [28]. In the case of bottle gourd, 10 to 70 per cent yield losses have been reported in Himachal Pradesh, India by, Jarial et al. [13]. The disease causes great havoc in pumpkin fields leading to up to 90 per cent yield loss [29]. It has also been reported from the pumpkin fields of Canada leading to 60 per cent yield losses [30].

4. Symptomatology

The disease has been reported to be prevalent on leaves and fruits of different cucurbits, such as watermelon, pumpkin, bottle gourd, cucumber and squashes. The symptoms of this malady have been described by various researchers on different crops. On Hubbard squash, the bacterial leaf spot was described to be mostly present on the lower surface of leaves [23]. These small water-soaked areas pierce the upper surface of the leaves as ill-defined yellow spots. These spots seem to grow in size to form a definite round spot of about 6 or 7 mm in width, even though they are restricted by veins, with a bright yellow halo prominent on the upper surface of the leaf. The lower surface of the leaf, however, shows spots with water-soaked margins or have no evident distinction. The bacterium has been known to enter through the stomata on leaves. As the infection spreads, large dead areas are produced by the coalescence, but these dead tissues do not drop out. Stems of young plants of the summer squashes have also been seen to be attacked by the pathogen. These young plants usually succumb to the injury caused to the growing points of the plant when it is infected. The symptoms of the disease on winter squash (Cucurbita moschata) [31] include fruit surface having circular water-soaked spots of 3 to 6 mm in width. These spots are somewhat sunken with the presence of sticky golden exudates.

These marks enlarge and became angular on ageing. Transverse section of the spot showed bacteria oozing out of the cut surface. It was later established that the symptoms on cucumber resemble that of angular leaf spots in features such as vein constraints, water-soaked spots on leaves with or in the absence of yellow halo and water-soaked injuries on fruits which may be deformed. The water-soaked spots on the leaves expand bit by bit to be later confined by the secondary veins [32].

On watermelon (Citrullus lanatus) leaves, the bacterium gives rise to angular and water-soaked spots with a chlorotic halo, which often becomes necrotic [21]. Along with this, scab-like lesions were also observed on fruits [20].

In case of bottle gourd, the disease appears on almost all plant parts of the crop [13]. Initially, small marginal chlorotic spots are formed on the margins of the leaves of any age group, which later merge and enlarge towards the centre of the leaf. In the later stages, these chlorotic spots convert into dead brown necrotic lesions and may cover the entire leaf lamina. These necrotic lesions do not drop as in the case of angular leaf spots. As the disease progresses, symptoms in the form of necrotic lesions may be seen developing on other plant parts such as stem/vine, tendrils and floral parts. In extreme cases, amber-coloured bacterial ooze is seen splitting the vine in two. In case of an attack on female flowers of the crop, the stigma and ovary rot, resulting in no fruit formation. On young fruits, the symptoms appear as water-soaked spots, which later lead to rotting of the entire fruit. On mature bottle gourd fruit, faint spots are visible, which later split to release amber-coloured bacterial ooze. Small, dark and angular lesions were observed on bottle gourd leaves, while the skin of the fruit showed soft, water-soaked dark green spots, which later changed into watery soft rot.

On pumpkin [26] foremost symptoms observed were small dark brown spots on the cotyledons itself. The margins of leaves of adult plants show chlorosis and necrotic spots, which later coalesce and finally give a dead brown colour appearance to the leaves. The necrotic areas on leaves do not fall, unlike angular leaf spots. The characteristic lesions on fruits begin with small oily, slightly depressed spots with a yellow border. The disease is also occasionally seen on stems along with fruits. In damp weather, amber-coloured ooze can be seen on the fruits. Lesions on pumpkin fruits have been described as scab-like [22]. On the foliage, water-soaked lesions are seen, which later turn brown with a prominent yellow halo. On enlarging, the lesions are observed to be limited by the veins, becoming angular. Pumpkin leaves develop small (1–2 mm) lesions, with indefinite yellow edges that may merge to form bigger necrotic areas usually on the leaf edges [17]. On pumpkin fruits, the appearance and types of symptoms may vary depending upon rind maturity and the presence of moisture. Initial lesions on fruits are small, sunken and circular spots ranging from 1/6 to 1/4 inch in diameter. These have a beige centre with dark brown halo. Later, the cuticle and epidermis break and the abrasion grows in size to be 1/2 inch in diameter. These huge lesions have a scab-like appearance and give rise to tan-coloured blisters. Bacterial leaf spot symptoms may develop on foliage as well as fruit. Small, round water-soaked symptoms can be seen on the underside of the leaves [33]. The corresponding upper surface of the leaf soon develops a yellow spot that later turns brown with a distinguished yellow halo. The leaf lesions usually remain small or may increase to about 7 mm in width. As they grow, they become angular, due to the constrictions offered by leaf veins. Another study outlined pumpkin lesions to have 1–4 mm irregular tan lesions with a yellow halo, while the matured ones displayed 1–4 mm tan-coloured sunken spots with dark brown borders, which later developed into severe soft rot [30]. Three different types of lesions were described on leaves of pumpkins based on the variation in colour and size. These ranged from small brown necrotic spots (1 mm) on leaves to angular spots (1–4 mm), with brown halo and beige centres. Small (1 mm), numerous, angular, tan-coloured necrotic spots on the leaves fall under Type 1 lesions. Spots with a beige centre and brown halo with an estimated size of 1–4 mm fall under Type 2 lesion category. Translucent angular spots with chlorotic halo, which are few in number on the leaf and measure up to 4 to 8 mm in diameter, fall under Type 3 lesions. Lesions on pumpkin, winter squash and gourd leaves are small up to 1.7 mm initially and enlarge to 1.7–3.81 mm in size [29]. Similarly, the circular spots on fruits are observed to be 1–2.5 mm initially, which enlarge to about 15.24 mm in diameter. In another study, symptoms on both upper and lower surfaces of the leaves were noticed but not on the fruit [25]. Lesions on the leaves are small, marginally sunken round spots with a beige centre and dark brown halo. On fruits, small, earthy-coloured spots appear, which later turn into necrotic spots attacked by saprophytes ultimately leading to fruit rot. On leaves, chlorotic spots were seen after rain storms in July, which later turned necrotic and coalesced (**Figure 1**) [23].

5. Causal organism

Bacterial spot of cucurbits is caused by Xanthomonas cucurbitae (ex. Bryan) Vauterin et al., 1995. Initially, this bacterium was named as Bacterium cucurbitae Bryan, Phytomonas cucurbitae (Bryan) Bergey et al., Pseudomonas cucurbitae (Bryan) Stapp, Xanthomonas cucurbitae (Bryan) Dowson, *Xanthomonas campestris* pv. cucurbitae (Bryan) Dye and is internally seed-borne in nature [34]. Gram staining test shows that bacterium is Gram-negative. Also, other tests show that it

Figure 1.
Symptoms of Xanthomonas cucurbitae on different host: (a) leaf, (b) twig, (c) flower, (d) stem and (e,f,g) fruit of bottle gourd. (h) Bacterial spot infected field area (i) leaves of sponge gourd (J,k) leaves and fruit of pumpkin.

is oxidase-negative and indole-negative, hydrolyses starch and esculin and forms pits on crystal violet pectate and carboxymethyl cellulose media [19]. Bacteria are rod-shaped and colonies are mucoid, circular, smooth textured, convex and glistening with entire margins and yellow in colour having a diameter of about 3–4 mm on nutrient agar medium. This bacterium hydrolyses esculin, which is clear from the blackening of esculin medium within 4 days of inoculation, digests protein which is confirmed as appearance of clear solution within 10 days of inoculation and liquefies gelatin as indicated by the appearance of a clear zone around bacterial growth [13].

6. Pathogenicity and host range

The disease is found and reported on many cucurbits throughout the world after its initial discovery [35]. Concentration of 10^5 to 10^8 cfu/ml induces diseases in different cucurbit hosts [28]. Incubation period of 3 to 5 days and 5 to 8 days has been reported on inoculated leaves and fruits, respectively, in the case of bottle gourd, cucumber, pumpkin and squash plants [13], while in inoculated leaves of pumpkin plants have been reported in 10 days [30]. Symptoms appear two days later on fruits in comparison to leaves [13]. T3SS helps to overcome host immunity to suppress MAMPs [36] and to manipulate host genes for infection [37].

7. Cultural studies

Cultural characteristics of bacterium were studied first by Bryan [17]. He described the bacterium as a short, rod-shaped, 0.5 to 1.3 × 0.45 to 0.6 μ in diameter with polar flagellum occurring singly in pairs or in short chains. Bacterium is non-acid fast staining, non spore producing and Gram negative. Colony characters are yellow and opalescent on beef agar. Minimum, maximum and optimum temperatures were found to be 4–6, 34–36, and 24–30°C, respectively, and a pH of 6.5 to 7 has been reported for colony development of X. cucurbitae isolates. [38]. For growth of bacteria, an optimum temperature range of 25 to 30°C and pH range of 6–7 has been reported in case of bottle gourd and pumpkin [28]. Regarding media, Yeast extract calcium carbonate broth has been reported the best followed by nutrient glucose broth for both bottle gourd and pumpkin while nutrient sodium chloride broth is for bottle gourd only [28]. As far as host range of the pathogen is concerned, the disease has been observed on almost all cucurbits including pumpkin, bottle gourd, cucumber, bitter gourd, sponge gourd, watermelon and other melons.

8. Molecular characterisation

Species identification was confirmed by using primers RST2 (5′AGGCCCT GGAAGGTGCCCTGGA3′) and RST3 (5′ATCGCACTGCGTACCGCGCGCGA3′) in PCR resulting in 1500 bp band. Identification of the pathogen was also identified by using the same primers resulting in 840-bp band. Also, amplification and sequencing of 16s rRNA gene resulted in 98 to 99 per cent similarity to X. cucurbitae accessions in GenBank [19]. Amplification of *hrpD* gene of X. cucurbitae in leaf spot disease of pumpkin was successfully carried out in isolates of *X. cucurbitae* [38]. PCR of X. cucurbitae with same primers resulted in 1.4 kb amplicon and gyrB rpoD fyuA dnaK resulted in isolate identity [23]. Genome sequencing and functional characterisation of cucurbitae studies lead to the first reference-quality whole-genome sequence of the X. cucurbitae, *which* isolates and showed that for infection in Pumpkin, both type II enzymes and type III effectors are necessary for infection [35].

9. Survival disease cycle and epidemiology

X. cucurbitae is internally seed borne [34], but exact location of the bacterium in the seed has not been located [27]. It has also been reported to survive on infected crop debris [33]. Fifty-three per cent of seed lots collected from pumpkin fields in Illinois were infected with X. cucurbitae [39]. The pathogen has been subsequently isolated from kernels and shells of naturally infected seeds. It also survives in infected leaves and fruit tissues of pumpkin for up to 24 months [40]. Pathogen is also known to survive in seed for more than 26 months at 4°C [41] and has been isolated from infected seeds even after 2 years of storage [42]. Much work has not been done on X. cucurbitae as yet, but available literature on other species of *Xanthomonas* reveals that seed is the most common method of survival for different *Xanthomonas* species.

The bacteria spread very rapidly in fields *via* rain splash in the soil. Fruit infection occurs through a natural opening or wound in young fruits prior to the development of a thick, waxy cuticle. Leaf infection, however, does not result in plant death of pumpkin, but likely plays a role in providing inoculum for fruit infection, as leaf infection occurs before fruit development. Therefore, control of leaf infection by X. cucurbitae may reduce fruit infection [20]. Long distance dispersal of pathogens is believed to be by contaminated seeds [34].

Disease development of bacterial spots of cucurbits occurs at temperatures ranging from 25–35°C and the pathogen stops spreading above 35°C [43]. The disease is favoured by warm humid weather and frequent rainfall [17, 42]. Temperatures between 25 and30°C with 90 per cent relative air humidity are favourable conditions for disease development [34]. Disease incidence reaches its peak at the end of July and the beginning of August [44].

Bacterial spots of cucurbits are affected by various epidemiological factors *viz.*, inoculation method, sowing of plant and plant age, which ultimately affect the initiation and development of infection. Among various methods of inoculation with isolates of the pathogen, syringe method of inoculation resulted in a minimum incubation period (2.33 and 3.33 days) and maximum disease severity (46.33 and 37.97%) in both bottle gourd and pumpkin, respectively, followed by pinprick method of inoculation. It was found that disease severity decreased with a delay in date of sowing from mid-May to the end of June (27th June) or mid-July (17th July) in pumpkin and bottle gourd [21]. A high disease severity was recorded in early sown (17 May and 7 June) crops [44]. Development of bacterial spots of bottle gourd and pumpkin was favoured by temperatures ranging between 30 and 35°C and relative humidity more than 80 per cent [21]. The plants of bottle gourd and pumpkin at the age of 10, 20 and 30 days were more susceptible to infection as compared to the older plants (40, 50 and 60 days old), which exhibited the presence of adult plant resistance in both the crops against X. cucurbitae [21]. It has been found that disease intensity continuously increases during vegetative phase and reaches its peak at the end of July or beginning of August [34]. Maximum temperature of 33.5°C and 85 per cent RH resulted in the increased disease intensity of bacterial spot of bottle gourd [32].

10. Host resistance

No X. cucurbitae-resistant cultivars have been identified to date. However, adult plant resistance, an enigmatic phenomenon in which resistance genes confer robust resistance at maturity but ineffective at seedling stage of plant [15], has been reported in two cultivars of cucumber *viz.*, collection 72–10 and Japanese Long Green which are moderately resistant against X. cucurbitae [36]. According to Babadoost, no cucurbit cultivar resistant to X. cucurbitae has been found yet [35]. Similar observations have been documented by Jarial et al. in case of bottle gourd where six different genotypes/varieties were observed for disease development and all were found to exhibit susceptible reactions towards the disease [45].

11. Management

Successful management of the disease needs integrated disease management to keep the disease below the economic threshold [46]. Several workers have mentioned various methods for disease management in affected regions of the world including cultural, seed treatments and foliar sprays. Crop rotation with non-cucurbitaceous crops for 2 years or more will decrease the disease levels [47]. Avoidance of overhead irrigation and working in fields when plants are wet (morning dew or after rain) reduces the bacterial spread from diseased to healthy plants [37]. Since the pathogen is seed-borne, disease management initiates with the use of pathogen-free seed [33]. Out of the various chemicals evaluated as seed treatments or foliar sprays against the pathogen, it was found that streptocycline, mancozeb, copper oxychloride, zineb and Bordeaux mixture were effective against the bacterium under *in vitro* experiments [16]. A seed dip treatment in a combination of streptocycline (100 ppm) plus copper oxychloride (3000 ppm) for 3.0 h has been reported to be quite effective in eliminating the bacterium from naturally infected seed [48].

Frequent foliar application of preventive sprays can help decrease bacterial spread in the field to some extent [37]. It has been reported that eight foliar sprays of chemicals such as plantomycin, paushamycin, streptocycline, Ceresan wet (phenyl mercury acetate), Blitox 50 (copper oxychloride) and captan were quite effective in managing the disease [39]. A management strategy comprising seed treatment with streptocycline (0.01%) plus copper oxychloride (0.3%) and four foliar sprays of the same combination at 10 days interval along with the removal of diseased plant parts regularly during the cropping season is a useful strategy against bacterial spot of bottle gourd [45]. In the field trials, copper oxychloride + copper hydroxide (Badge X2 DF), copper sulphate (Cuprofix Ultra 40 DF), oxytetracycline (Mycoshield 40 WSP), copper sulfate pentahydrate (Phyton-016B), copper hydroxide (Kocide-3000 46.1 DF) plus acibenzolar-s-methyl (ActiGard 50 WG), Kocide-3000 46.1 DF plus famoxadone + cymoxanil (Tanos 50D WG), an extract from Reynoutria sachalinensis (Regalia) and *B. subtilis* (Serenade ASO) were more effective in reducing incidence and severity of bacterial spot on both leaves and fruit compared to controls [48]. These chemical compounds or biocontrol agents may be used in combination with other methods to manage X. cucurbitae in pumpkins. It was found that a chemical combination comprising streptocycline (100 ppm) + captan (2500 ppm) proved effective in controlling the pathogen and disease both under *in vitro* and pot house conditions [49].

12. Conclusions

The gourd family or *Cucurbitaceae* is the second largest family among fruits and vegetables next to Solanaceae and contains edible crops such as cucumber, pumpkin and melon. Bacterial leaf spot is caused by *X. cucurbitae* and was reported for the first time on Hubbard squash in New York in 1926. Initially, the bacterium was named Bacterium cucurbitae by Bryan, later Phytomonas cucurbitae (Bryan) Bergey et al., Pseudomonas cucurbitae (Bryan) Stapp, Xanthomonas cucurbitae (Bryan) Dowson, *Xanthomonas campestris* pv. cucurbitae (Bryan) Dye. It is one of the most important diseases of cucurbits such as watermelon, pumpkin, bottle gourd, cucumber and squashes in the world and leads to huge losses of up to 100 per cent. The disease has been reported to be prevalent on leaves and fruits of different cucurbits, such as watermelon, pumpkin, bottle gourd, cucumber and squashes. The bacterium has been known to enter through the stomata on leaves. As the infection spreads, large dead areas are produced by the coalescence, but these dead tissues do not drop out. Stems of young plants of the summer squashes have also been seen to be attacked by the pathogen. These young plants usually succumb to the injury caused to the growing points of the plant when it is infected. This disease is favoured by warm humid weather and frequent rainfall. Bacteria is Gram negative, rod shaped. Colonies are yellow, mucoid, circular, smooth textured, convex and entire margins are glistening with diameter of about 3–4 mm diameter, oxidase-negative, indole-negative; hydrolyzes starch and esculin and forms pits on crystal violet pectate and carboxymethyl cellulose media. Minimum, maximum and optimum temperatures were found to be 4–6, 34–36, and 24–30°C, respectively, and a pH of 6.5 to 7 has been reported for colony development of X. cucurbitae isolates. Pathogen is internally seed-borne and also perpetuates in infected crop debris. Pathogen is also known to survive in seed for more than 26 months at 4°C and has been isolated from infected seeds even after 2 years of storage. Much work has not been done on X. cucurbitae as yet, but available literature on other species of *Xanthomonas* reveals that seed is the most common method of survival for different *Xanthomonas* species. No X. cucurbitae-resistant cultivars have been identified to date. Since the pathogen is seed-borne, disease management initiates with the use of pathogen-free seed. Crop rotation with non-cucurbitaceous crops for 2 years or more will decrease the disease levels. However, Streptocycline reduces the disease severity up to some extent in the field. Current research is being conducted on molecular characterisation of different isolates and possible management strategies along with antibiotics.

Conflict of interest

The authors declare no conflict of interest.

Author details

Kumud Jarial[1*], Rajender Singh Jarial[1], Priyanka Bhardwaj[1], Arpana Sharma[2] and Somya Hallan[3]

1 College of Horticulture and Forestry, Dr. Y.S. Parmar University of Horticulture and Forestry, Neri Hamirpur, H.P., India

2 Punjab Agricultural University, Ludhiana, Punjab, India

3 CSK Himachal Pradesh Krishi Vishwavidyalaya, Palampur, H.P., India

*Address all correspondence to: kumudvjarial@rediffmail.com

References

[1] Van Laethem S, Frans M, Aerts R. pH modulation of the environment by *Stagonosporopsis cucurbitacearum*, an important pathogen causing fruit rot in Cucurbitaceae. European Journal of Plant Pathology. 2021;**159**:235-245. DOI: 10.1007/s10658-020-02164-w

[2] Schaefer H, Renner SS. Phylogenetic relationships in the order Cucurbitales and a new classification of the gourd family (Cucurbitaceae). Taxon. 2011;**60**(1):122-138

[3] Saffanah MAA, Rosli WWI. Therapeutic benefits of commercially available gourd family in improvement and sustainability of human health. International Journal of Engineering & Technology. 2018;**7**:164-166

[4] Salehi B, Quispe C, Sharifi-Rad J, Giri L, Suyal R, Jugran AK, et al. Antioxidant potential of family Cucurbitaceae with special emphasis on Cucurbita genus: A key to alleviate oxidative stress-mediated disorders. Phytotherapy Research. 2021;**35**:3533-3557

[5] Chomicki G, Hanno Schaefer H, Renner SS. Origin and domestication of Cucurbitaceae crops: Insights from phylogenies, genomics and archaeology. New Phytologist. 2019;**226**:1240-1255

[6] Ahmad G, Khan AA. Pumpkin: Horticultural importance and its roles in various forms; a review. International Journal of Horticulture and Agriculture. 2019;**4**(1):1-6

[7] Lestari B, Meiyanto E. A review: Emerging nutraceutical potential of pumpkin seeds. Indonesian Journal of Cancer Chemoprevention. 2018;**9**(2):92-101

[8] Chandramouli B, Reddy RVSK, Babu MR, Jyothi KU, Umakrishna K, Rao MP. Genetic variability studies for yield and yield attributing traits in F2 generation of bottle gourd (*Lagenaria siceraria* (Mol.) Standl.). The Pharma Innovation Journal. 2021;**10**(5):1484-1488

[9] Khan A, Hussain Q, Asim M, Khan N, Habibullah UR, Ali M, et al. Evaluation of bitter gourd varieties on different methods of cultivation. Pure and Applied Biology. 2020;**11**(1):58-71. DOI: 10.19045/bspab.2022.110007

[10] Adeoye I, Olajide D, Adeb A, Usman J, Badmus M. Economic analysis of watermelon based production system in Oyo state. Journal of Agricultural and Biological Sciences. 2011;**6**(7):53-59

[11] Joshi BK, Han BKC, Tiwari RK, Ghale M, Sthapit BR, Upadhyay MP. Descriptors for Sponge Gourd Luffa Cylindrica (L.). Roem. NARC, LIBIRD and IPGRI; 2004

[12] Tatlioglu T. Cucumber (Cucumis sativus L.). In: Kailov G, Bergn B, editors. Genetic Improvement of Vegetable Crops. Oxford Pergamon Press; 1997. pp. 197-227

[13] Jarial K, Dogra BS, Mandradia RK, Kumar S, Sharma D, Gupta AK. Investigations on a new bacterial disease of bottle gourd in sub-tropical zone of Himachal Pradesh. Plant Disease Research. 2011;**26**(1):68-75

[14] Liu Q, Ravanlou A, Babadoost M. Occurrence of bacterial spot-on pumpkin and squash fruit in north central region of United States and bacteria associated with the spots. Plant Disease.

2016;**100**(12):2377-2382. DOI: 10.1094/PDIS-01-16-0107-RE

[15] Sharma A, Jarial K, Jarial RS. Effect of plant age on the bacterial spot of bottle gourd and pumpkin caused by *Xanthomonas cucurbitae*. International Journal of Current Microbiology and Applied Sciences. 2020;**9**(12):2205-2213. DOI: 10.20546/ijcmas.2020.912.260

[16] Sulley S, Huang Y, Hind SR, Babadoost M. Screening and identification of Cucurbita germplasm resistant to *Xanthomonas cucurbitae*, incitant of cucurbit bacterial spot. Plant Pathology. 2021;**70**(9):2188-2196. DOI: 10.1111/ppa.13445

[17] Bryan MK. Bacterial leaf spot of squash. Journal of Agricultural Research. 1930;**40**(4):385-391

[18] Vlasov V V. 2005. *Xanthomonas cucurbitae* - causal agent of bacteriosis on cucurbits in Pridnestrov.e. In: Collection of Papers of Participants of the International Scientific Conference. Phytopathogenic Bacteria. Phytoncidology. Allelopathy. by VS Podgorskii (ed.), Kiev: Gosudarstvennyi argoekologicheskii universitet; 2005. p. 14-18

[19] Dutta B, Gitaitis RD, Lewis KJ, Langston DB. A new report of *Xanthomonas cucurbitae* causing bacterial leaf spot of watermelon in Georgia, USA. Plant Disease. 2013;**97**:556

[20] William RH, Zitter TA. Bacterial Leaf Spot: Compendium of Cucurbits Diseases. St. Paul MN: American phytopathological Society; 1996, 1996. p. 35

[21] Pruvost O, Robène-Soustrade I, Ah-You N, Jouen E, Boyer C, Wuster G, et al. First report of *Xanthomonas cucurbitae* causing bacterial leaf spot of watermelon in the Seychelles. Plant Disease. 2009;**93**(6):671

[22] Lamichhane JR, Varvaro L, Balestra GM. Bacterial leaf spot caused by *Xanthomonas cucurbitae* reported on pumpkin in Nepal. New Disease Reports. 2010;**22**:20

[23] Altin I, Casoli L, Stefani E. First report of bacterial spot caused by *Xanthomonas cucurbitae* on pumpkin in Italy. New Disease Reports. 2020;**41**:21. DOI: 10.5197/j.2044-0588.2020.041.021

[24] Sharma RR. Integrated Management of Bacterial Spot of Bottle Gourd Caused by *Xanthomonas cucurbitae* [thesis]. Chatha, Jammu: Sher-e-Kashmir University of Agricultural Sciences & Technology of Jammu Main Campus; 2016

[25] Chinnadurai C, Ramkisson A, Ramjegatheseh DS, Ramsubhag A, Jayaraman. Disease status of pumpkin in the southern Caribbean. Tropical Agriculture. 2018;**95**(2):36-41

[26] Larazev AM. Diseases: *Xanthomonas camspestris* pv.*Cucurbitae* (Bryan) Dye - bacterial leaf spot of cucurbits. In: 2003-2009 Project "Interactive agricultural ecological Atlas of Russia and neighbouring countries: Economic plants and their diseases, pests and weeds". 2009. Available from: http://www.agroatlas.ru/en/content/diseases/Cucurbitae/Cucurbitae_Xanthomonas_campestris_pv_cucurbitae/

[27] Babadoost M. Disease Management of Cucurbits. USA: University of Illinois; 2013

[28] Kumari S, Jarial K, Jarial RS, Jandaik S. Cultural parameters of *Xanthomonas cucurbitae* causing bacterial leaf spot of bottle gourd and

pumpkin. International Journal of Bio-resource and Stress Management. 2020;**11**(3):214-218

[29] Salamanca LR. Bacterial Diseases of pumpkins: An old enemy and an emerging bacterial disease. Michigan State University Extension Bulletin. 2014. Available from: http://msue.anr.msu.edu

[30] Trueman CL, Roddy E, Goodwin PH. First report of bacterial spot (*Xanthomonas cucurbitae*) of pumpkin in Ontario, Canada. New Disease Reports. 2014;**30**:8. DOI: 10.5197/j.2044-0588.2014.030.008

[31] Laf JG, Linardelli CE, et al. Detection of *Xanthomonas cucurbitae* (Bryan) Vauterin affecting fruits of winter squash in Mendoza and San Juan, Argentina. Revista de la Facultad de CienciasAgrarias. 2011;**43**(2):245-253

[32] Lia RS, Md AR, Akter A, Mst SSE, Md HF, Acharjee UK, et al. Detection of *Xanthomonas campestris* pv. *Cucurbitae* from bacterial leaf spot disease of cucumber and evaluation of its biological control. Advances in Bioresearch. 2018, 2018;**9**:41-46

[33] Goldberg N. Bacterial leaf spot of cucurbits fact sheet, Plant Pathology: The news you can use. New Mexico State University Extension Bulletin. Oct 2012;**2012**. Available from: http://aces.nmsu.edu

[34] Vincent-Sealy L, Brathwaite CWD. Bacterial leaf spot of cucumber in Trinidad. Tropical Agriculture. 1982;**59**(4):287-288

[35] Rai R, Pasion J, Majumdar T, Green C, Hind S. Genome sequencing and functional characterization of *Xanthomonas cucurbitae*, the causal agent of bacterial spot disease of cucurbits. Phytopathology. 2021;**111**(8):1289-1300

[36] Buttner D, Bonas U. Regulation and secretion of Xanthomonas virulence factors. FEMS Microbiology Reviews. 2010;**34**:107-133. DOI: 10.1111/j.1574-6976. 2009. 00192.x

[37] Buttner D. Behind the lines–actions of bacterial type III effector proteins in plant cells. FEMS Microbiology Reviews. 2016;**40**(8):94-937. DOI: 10.1093/femsre/fuw026

[38] Hasan Zia SM, Md H, Zannati Z, Md H, Md I, Md K, et al. PCR amplification of DNA sequence related to the hrpD gene of *Xanthomonas cucurbitae* in leaf spot disease of pumpkin and their antagonism by soil bacteria. Archives of Phytopathology and Plant Protection. 2018;**51**:1-15. DOI: 10.1080/03235408.2018.1460923

[39] Ravanlou A. Occurrence of Bacterial Spot of Pumpkin (*Xanthomonas cucurbitae*) in Illinois and Pathogenic and Genetic Variations, Host Range, and Seed Borne Aspect of the Pathogen [thesis]. USA: University of Illinois at Urbana-Champaign; 2013

[40] Thapa S. Field Survival of *Xanthomonas cucurbitae*, the Causal Agent of Bacterial Spot of Pumpkin, and Efficacy of Selected Chemicals and Biocontrol Agents for Control of the Diseases [thesis]. USA: University of Illionois at Urbana-Campaign; 2014

[41] Zhang X. Determining the Characteristics of *Xanthomonas cucurbitae* Bacterium from the North Central Region and Developing Effective Seed Treatment for Eradication of the Bacterium in Pumpkin Seeds [thesis]. USA: University of Illinois at Urabana-Campaign; 2016

[42] ZhangX,BabadoostM.Characteristics of Xanthomonas cucurbitae isolates from pumpkins and survival of the bacterium

in pumpkin seeds. Plant Disease. 2018;**102**(9):1779-1784

[43] Bineeta S B, Majuinder S, Kumar S. Fungal and bacterial diseases. In Disease of Horticultural Crops, Vegetables, Ornamentals and Mushrooms by L. R. Verma and R. C. Sharma (eds), Indus Publishing Cooperation, New Delhi, India; 1999. Pp. 251-252

[44] Sharma A, Jarial K, Jarial RS, Sharma D. Effect of sowing dates on the bacterial leaf spot of bottle gourd and pumpkin caused by *Xanthomonas cucurbitae*. International Journal of Bio-resource and Stress Management. 2020;**13**:106-113

[45] Jarial K, Jarial RS, Sharma D, Dogra BS, Kumar S. Strategy for the management of bacterial spot of bottle gourd caused by *Xanthomonas cucurbitae* under low hill conditions of Himachal Pradesh. International Journal of Bioresource and Stress Management. 2015;**6**:744-748

[46] Babadoost M, Zitter TA. Fruit rots of pumpkin a serious threat to the pumpkin industry. Plant Disease. 2009;**93**:772-782

[47] Anonymous. Bacterial Spot of Cucurbits. USA, No. 949: University of Illinois Extension Report on Plant Disease; 2012

[48] Thapa S, Babadoost M. Effectiveness of chemical compounds and biocontrol agents for management of bacterial spot of pumpkin caused by *Xanthomonas cucurbitae*. Plant Health Progress. 2016;**17**:106-113

[49] Sharma A. Studies on Management of Bacterial Leaf Spot of Bottle Gourd and Pumpkin Caused by Xanthomonas cucurbitae [thesis]. Nauni, Solan: Dr. Y. S Parmar University of Horticulture and Forestry; 2020

Chapter 4

Biological Control of Diseases of Bottle Gourd

Efath Shahnaz, Saba Banday, Ali Anwar, Qadrul Nisa, Gazala Gulzar, Atufa Ashraf and Diksha Banal

Abstract

Biological control of plant diseases is an important component of disease management, particularly in the today's' world of environmental consciousness and awareness. It is particularly preferred method of disease management under organic production system. Biological control is successful in almost all the crops against a number of diseases but soil borne diseases are most responsive to bio-control methods. The agents of biological control, known as bio-control agents (BCAs) belong to a vast group of micro-organisms, particularly fungi (*Trichoderma, Ampelomyces*, etc), bacteria (*Pseudomonas, Bacillus*, etc) and actinomycetes. Bottle gourd is an important vegetable crop belonging to the family *Cucurbitaceae*. It suffers from a number of diseases like anthracnose, powdery mildew, downy mildew, wilt, etc. The present review shall be an attempt to review the biological control of the major diseases of bottle gourd.

Keywords: biological control, trichoderma, bottle gourd, disease management, mechanism of biocontrol

1. Introduction

Fungicides have been used for the management of plant diseases, since the time man became interested in diseases of plants, besides invoking gods for protecting them from the wrath of diseases. However, the dawn of century saw an increasing awareness and consciousness relating to the adverse effects of these chemicals, particularly towards our environment. As a result there is advanced advocacy against the use of fungicides and search for alternative ways of disease management. Resistance to plant diseases is the most preferred way but most of the times not practical due to obvious and inherent reasons of disease resistance. Cultural practices have also been used since times immemorial, often times without the active consciousness of the producer. However, relying on cultural practices alone is not feasible. More so in case of diseases with compound interest type of growth. Biological control has immense potential to be used as viable and alternative disease management strategy and it can be blended well with the integrated disease management capsules [1–5]. Biological control has proved very effective in disease management, more so of soil borne diseases [6–9], diseases of fruits [10–12], foliar diseases [13–18] as well as nematode diseases [14, 19, 20], besides innumerable other crop diseases and the list is growing day by day.

IntechOpen

Bottle gourd is an important crop grown all over the world for its culinary and medicinal properties [21]. However, it is affected by a number of diseases like anthracnose, downy mildew, powdery mildew, bacterial leaf spot, mosaic, etc., all of which lead to severe constraints in yield and reduction in the realization of full genetic potential. Management has been attempted through use of chemicals [22–25], plant extracts [26, 27], resistance [28–30] or integrated disease management [31–34]. However, in the recent past focus has shifted on the biological control of plant diseases and bottle gourd is no exception. In some cases, remarkable success has been obtained through the use of biocontrol agents, while in some cases triumphs have been restricted to lab studies. Following is an attempt to review biological control of diseases of bottle gourd in the last few decades.

2. Fungi as biocontrol agents of diseases of bottle gourd

A number of fungi have been used efficiently against different diseases of bottle gourd. Most commonly used fungi are different species of *Trichoderma.* These fungi have versatile nature, are easy to isolate from native soils and can easily be cultured, as a result of which many studies have focused on use of *Trichoderma* as bio-control agents. Also, most of the *Trichoderma* spp. employ diverse mechanisms for disease control like mycoparasitism, competition antibiosis, as well as induction of systemic resistance [35, 36].

T. harzianum reduced the mycelial growth of *Fusarium moniliforme*, the causal agent of bottle gourd wilt [37]. Although most of the microbial antagonists viz., *Trichoderma harzianum, T. viride, Gliocladium virens, Bacillus subtilis* and *Stachybotrys atra* significantly reduced seedling mortality and root rot infection of *F. oxysporum* in bottle gourd and cucumber, *T. harzianum* was found most effective [38]. In Gujarat, India, the isolate of *T. viride* (Sardarkrushinagar) was most effective against *F. oxysporum* followed by*y T. harzianum* (Junagadh), *viride* (Junagadh), *T. viride* (Navsari), *Bacillus subtilis* (Sardarkrushinagar) *and Pseudomonas fluorescens* (Sardarkrushinagar) [39]. Besides being healthy and free from wilt symptoms, the pathogen quantum inside the host and soil was reduced in bottle gourd seedlings raised from antagonist coated seeds [40]. Besides *Trichoderma, Penicillium citrinum* and *Aspergillus flavus* appear to effectively reduce mycelial growth of *F. solani* [41].

The causal organism of anthracnose of bottle gourd, viz. *Colletotrichum lagenarium* was inhibited by *T. viride* [42].

T. hamatum was found more promising than *T. harzianum,* although both inhibited the growth and sporulation of *Alternaria alternata,* the causal organism of black rot of bottle gourd [43].

Chatur and Anil (2014) reported that seed treatment and soil application of *Trichoderma harzianum and T. viride* were effective for the management of gummy stem blight and these treatments also increased fruit yields [44]. Patel *et al.* (2017) reported similar results against the same disease using same bio-agents and reported that next in order were *Bacillus subtilis* and *Pseudomonas fluorescence* [45]. Soil application of spent mushroom substrate enriched with *T. harzianum* significantly increased yield and simultaneously decreased disease incidence of gummosis [46]. *Gliocladium virens* was found most effective in reduction of seed and root infection caused by *Lasiodiplodia theobromae,* whereas, *B. subtilis* helped in the reduction of seed and seedling infection of bottle gourd under *in vitro* conditions [47].

Seed priming by aqueous solutions of culture filtrates of *Trichoderma* can be used for vegetable seed treatment, including bottle gourd, for controlling seed-borne fungal infection [48]. Seed treatment with *T. harzianum* resulted in the induction of systemic resistance [49]. Besides, *Trichoderma* increases the germination percentage of treated seeds [50]. As a result of these studies, it has been recommended that seed treatment with *Trichoderma,* may be included in the IPM schedule to increase the net return of farmers [51]. *Trichoderma* enriched bio-fertilizers promote crop cultivation of bottle gourd with subsequent reduction in the usage of nitrogen fertilizers [52].

The fungus *Paecilomyces lilacinus* had better rate of success against second stage juveniles of root knot nematode infesting bottle gourd than either *Trichoderma* or *Pseudomonas* [53]. Tricho-compost was effective in reducing the root knot severity and increasing plant growth and yield of bottle gourd [54].

3. Bacteria as biocontrol agents of diseases of bottle gourd

In Japan, the root systems of an associate crop of welsh onion or Chinese chive were dipped in broth cultures of *Pseudomonas gladioli* and then bottle gourd was mixed cropped with the associate crop. It was observed that by this treatment, the occurrence of Fusarium wilt was suppressed to a large extent [55].

Pseudomonas fluorescens as foliar spray has been used for the management of Alternaria leaf blight of bottle gourd [24]. Along with some other fungal biocontrol agents, particularly *Trichoderma,* bacterial biocontrol agents like *Bacillus subtilis* controlled the pre- and post-emergence infection of *Lasiodiplodia theobromae* in seedlings of bottle gourd under both in vitro and in vivo conditions [47]. In another experiment on the efficacy of biocontrol agents against gummy stem blight of bottle gourd, it was found that although the bacterial biocontrol agents were not much effective against the disease, but treatment with *Bacillus subtilis* and *Pseudomonas fluorescens* did increase the fruit yield over control [45]. Combined treatment of bottle gourd with carbendazim (seed treatment), mancozeb (foliar spray) and *Pseudomonas fluorescens* (foliar spray) resulted in minimum disease incidence and minimum disease severity of *Alternaria* blight with maximum disease control [24].

A study by Rani et al. (2022) revealed that bacteria like *Bacillus amyloquefaciens*, *B. megaterium*, *P. fluorescens*, and *P. putida*, could be used for the management of root knot nematode both under in vivo and in vitro conditions [56]. Seed treatment with *Bacillus subtilis* reslted in the reduction of seed borne fungi of vegetable seeds including bottle gourd [57]. *Pseudomonas fluorescens* was the second best bioagent after *Paecilomyces lilacinus* for causing the most mortality of second stage juveniles of root knot nematodes [53].

4. Mechanism of biocontrol

The biocontrol agents utilize a variety of mechanisms for reducing the pathogen populations and promote plant growth. The general mechanisms of biocontrol agents have been reviewed by a number of scientists [58, 59]. Most of the mechanisms can be categorized as (a) antibiosis, (b) competition for food and space, (c) hyperparasitism or mycoparasitism, (d) cell wall degrading enzymes and (e) induction of systemic resistance. The mechanisms of action of BCAs have been demonstrated using microbiological, microscopic and biochemical techniques. In the recent past, development and use of molecular techniques have yielded significant results [60].

In some cases, biocontrol efficacy can be increased by the use of more than one biocontrol agent with more than one mode of action. An example of this situation is the use of a yeast (*Pichia guilermondii*) which caused inhibition of conidial germination of *Botrytis cinerea,* and the bacteria (*Bacillus mycoides*) that caused breakdown and destruction of conidia [61]. However, we have to tread this path with caution because it has been found that in combined use of BCAs, antagonistic interactions among BCAs are more likely to occur than synergistic interactions [62]. For example, it has been found that DAPG from *Pseudomonas fluorescens* strains enhanced *nag1* N-acetyl-β-d-glucosaminidase, but not ech42 endochitinase expression, whereas an unknown substance from *P. fluorescens* CHA0 repressed expression of both *Trichoderma* chitinases [63].

The production of plant growth promoting metabolites as well as antagonistic potential of different BCAs varies with respect to disease control of bottle gourd is expected. Kotasthane *et al.* (2015) found that the production of metabolites in 20 different isolates of *Trichoderma* did not corelate with enhanced growth on cucumber, bottle gourd and bitter gourd [64]. The isolate viz. *T. viride* isolate (T14) was identified as highest producer of inorganic phosphate, IAA and siderophore and exhibited high antagonistic and plant growth promoting ability. *T. harzianum* strain T-A66 promoted growth of bitter gourd and induced disease resistance to *Fusarium oxysporum* by inducing quick H_2O_2 burst and callose deposition, as well as increasing antioxidant enzyme activities and phenolic compounds content [65]. Munir et al. (2019) demonstrated the inhibitory role of chitinolytic enzyme extracts of *Trichoderma* against fungal pathogens of bottle gourd, whereas, Shah et al., showed the inhibitory effect of various *Trichoderma* isolates on the mycelia of pathogens of bottle gourd [66, 67].

In another study on powdery mildew of *Cucurbitaceae*, it was found that the antagonistic strain of *Bacillus subtilis* confers protection against cucurbit powdery mildew by the production of reactive oxygen species and cell wall reinforcement and by activation of jasmonate and salicylic dependent responses [68]. The scanning electron micrographs revealed that the antagonistic bacteria colonized the leaves by forming orderly microcolonies following epidermal cell junctions and were closely attached to *Podosphaera fusca* conidia and hyphae resulting in the collapse of latter [69]. The lipopeptides produced by *B. subtilis* are also able to reduce the disease by arresting conidial germination, probably due to induction of cytological alterations [70]. Similarly, the epiphytic yeast *Pseudozyma aphidis* proliferated on the infected tissue and its long hyphae parasitized the powdery mildew hyphae and spores as an ectoparasite, besides producing antibiotics [71].

Ongena *et al.* (1999) suggested that antifungal compounds induced by inoculation of cucumber roots with fluorescent *Pseudomonas* sytrains protected the cucumber plants against *Pythium aphanidermatum* and siderophores or antibiosis had minimal role in protection against disease [72].

In case of bacterial diseases compounds like iturin like lipopeptides have been found to be of considerable importance. The antibacterial activity is absent in iturin deficient mutants. Fluoresence and transmission electron microscopic studies have revealed that these compounds are cytotoxic to the bacterial plasma membrane [73]. Iturins have also been implicated in the antagonism of *B. subtilis* towards *P. fusca* [74].

5. Conclusion

Biological control has minimized the safety concerns regrading chemical fungicides and pesticides. Bio-control agents do not leave harmful residues in soil, are long lasting, environment friendly and self-sustaining in the long run. The problem of disease resistance is minimized and most of the biocontrol agents have plant growth promoting properties as well. However, they can be less effective than chemicals, particularly when time has to be taken into consideration. Although biological control proves to be cheaper in the long run, the initial investment and startup production costs limits its Commercialization and formulation of successful biocontrol agents pose another problem. There is the problem of culturing biocontrol agents in large quantities. They cannot be stored for long periods of time and hence have low shelf life. Moreover, there is very little evidence that biological control in itself is sufficient for practical management of plant diseases. In most of the cases, biological control needs to be integrated with other methods of disease control for effective management of diseases. It also has to be planned well in advance and is not efficient strategy for mitigation of emergencies like outbreaks of blights where the disease progresses at an exponential rate.

Biological control is the latest and most interesting strategy that is being used in the management of diseases of plants including bottle gourd. However, at present very little literature is available on the biological control of different diseases of bottle gourd and the work on plant growth promoting microbes is scantier. The need of the hour is to accelerate work on this aspect and search for BCAs with excellent biocontrol potential against maximum diseases of bottle gourd with bio-stimulatory action as well. These microbes need to be formulated and commercialized for the benefit of mankind in general and farmers in particular.

Author details

Efath Shahnaz[1*], Saba Banday[2], Ali Anwar[3], Qadrul Nisa[2], Gazala Gulzar[2], Atufa Ashraf[2] and Diksha Banal[2]

1 Dryland Agricultural Research Station, Rangreth, SKUAST-Kashmir, India

2 Faculty of Horticulture, Division of Plant Pathology, SKUAST-Kashmir, India

3 Faculty of Agriculture, Division of Plant Pathology, Wadura, SKUAST-Kashmir, Jammu & Kashmir, India

*Address all correspondence to: efaths@gmail.com

References

[1] Anwar A, Efath Shanaz MN, Mughal SB, Bashir T. Integrated disease management of downy mildew disease in bottle gourd (*Lagenaria siceraria* (Mol.) Standl) under temperate conditions of Kashmir. International Journal of Current Microbiology and Applied Sciences. 2021;**10**(03):824-828. DOI: 10.20546/ijcmas.2021.1003.104

[2] Efath Shahnaz VK, Razdan SEH, Rizvi TR, Rather SG, Andrabi M. Integrated disease management of foliar blight disease of onion: A case study of application of confounded factorials. Journal of Agricultural Science. 2013;**5**(1):17-22

[3] Rather TR, Razdan VK, Tewari AK, Shahnaz E, Bhat ZA, Mir GH, et al. Integrated management of wilt complex disease in bell pepper (*Capsicum annum* L.). Journal of Agricultural Science. 2012;**4**(7):141-147

[4] Ons L, Bylemans D, Thevissen K, Cammue BP. Combining biocontrol agents with chemical fungicides for integrated plant fungal disease control. Microorganisms. 2020;**8**(12):1930

[5] Yadav R, Bunker RN, Sharma SS, Trivedi A, Rawal P. Survey, incidence and integrated disease management of cotton root rot caused by *Rhizoctonia solani* (Kuhn.). The Pharma Innovation Journal. 2022;**11**(8):1618-1621

[6] Bubici G, Kaushal M, Prigigallo MI, Gómez-Lama Cabanás C, Mercado-Blanco J. Biological control agents against *Fusarium* wilt of banana. Frontiers in Microbiology. 2019;**10**:616

[7] Bhusal B, Mmbaga MT. Biological control of *Phytophthora* blight and growth promotion in sweet pepper by *Bacillus* species. Biological Control. 2020;**150**:104373

[8] Niu B, Wang W, Yuan Z, Sederoff RR, Sederoff H, Chiang VL, et al. Microbial interactions within multiple-strain biological control agents impact soil-borne plant disease. Frontiers in Microbiology. 2020;**11**:585404

[9] Nelson EB, Burpee LL, Lawton MB. Biological control of turfgrass diseases. In: Handbook of Integrated Pest Management for Turf and Ornamentals. London: CRC Press; 2020. pp. 409-424

[10] Janisiewicz W. Biological control of diseases of fruits. In: Biocontrol of Plant Diseases. Boca Raton: CRC Press; 2020. pp. 153-165

[11] Moraes Bazioli J, Belinato JR, Costa JH, Akiyama DY, Pontes JG, Kupper KC, et al. Biological control of citrus postharvest phytopathogens. Toxins. 2019;**11**(8):460

[12] Leneveu-Jenvrin C, Charles F, Barba FJ, Remize F. Role of biological control agents and physical treatments in maintaining the quality of fresh and minimally-processed fruit and vegetables. Critical Reviews in Food Science and Nutrition. 2020;**60**(17):2837-2855

[13] Banoo A, Shahnaz E, Banday S, Rasool R, Bashir T, Latif R. Studies on predominant epiphytic micro-flora as antagonists to post-harvest pathogens of apple. Journal of Plant Pathology and Microbiology. 2020;**11**:512

[14] Anwar A, Shahnaz E. Sustainable induction of systemic resistance in response to potential biological control agents in crops. In: New and

Future Developments in Microbial Biotechnology and Bioengineering. Netherlands, United Kingdom, United States: Elsevier; 2022. pp. 151-176

[15] Shahnaz E, Razdan VK, Banday S. Evaluation of bacterial bio-control agents against foliar blight pathogens of onion in Jammu region. SKUAST Journal of Research. 2016;**18**(1):12-15

[16] Amin F, Qazi NA, Banday S, Dar SH, Shahnaz E. Biological control of powdery mildew and black spot diseases of rose. Journal of Pharmacognosy and Phytochemistry. 2018;**7**(3):2826-2828

[17] Baccari C, Antonova E, Lindow S. Biological control of Pierce's disease of grape by an endophytic bacterium. Phytopathology. 2019;**109**(2):248-256

[18] Wonglom P, Daengsuwan W, Ito SI, Sunpapao A. Biological control of *Sclerotium* fruit rot of snake fruit and stem rot of lettuce by *Trichoderma* sp. T76-12/2 and the mechanisms involved. Physiological and Molecular Plant Pathology. 2019;**107**:1-7

[19] Ahmad G, Khan A, Khan AA, Ali A, Mohhamad HI. Biological control: A novel strategy for the control of the plant parasitic nematodes. Antonie Van Leeuwenhoek. 2021;**114**(7):885-912

[20] Abd-Elgawad MM. Biological control agents in the integrated nematode management of potato in Egypt. Egyptian Journal of Biological Pest Control. 2020;**30**:1-3

[21] Zahoor M, Ikram M, Nazir N, Naz S, Batiha GE, Kamran AW, et al. A comprehensive review on the medicinal importance; biological and therapeutic efficacy of *Lagenaria siceraria* (Mol.) (bottle gourd) Standley fruit. Current Topics in Medicinal Chemistry. USA. 2021;**21**(20):1788-1803

[22] Sharmin S. Management of powdery mildew disease of bottle gourd (*lagenaria siceraria*) using potassium and sulphur containing modern phyto-chemicals (Doctoral dissertation)

[23] Singh G, Jain S. Field and In Vitro effectiveness of fungicides against gummy stem disease, sudden death and leaf blight, the emerging syndromes of bottle gourd. Pesticide Research Journal. 2020;**32**(2):316-325

[24] Maheshwari SK, Choudhary BR, Sharma BD, Saroj PL. Management of *Alternaria* leaf blight of bottle gourd in Western Rajasthan, India. International Journal of Current Microbiology and Applied Sciences. 2017;**6**(6):1272-1277

[25] Chauhan RS, Bhatia JN. Evaluation of disease control potentiality of certain fungicides in controlling anthracnose disease of bottle gourd. Plant Disease Research. 2012;**27**(2):237-238

[26] Yaseen S. Analysis of anti-fungal activity of selected medicinal plant extracts against Alternaria leaf blight of bottle gourd (Doctoral dissertation, University of Agriculture, Faisalabad). 2020

[27] Joshi S, Borkar P, Joshi M. In vitro assessment of bio-agents, phyto extracts/organicals and fungicides against *Colletotrichum lagenarium*. Trends in Biosciences. 2019;**12**(20):1321-1326

[28] Soni N, Raj K, Chugh DD. Screening of bottle gourd [Lagenaria siceraria (Molina) Standl.] germplasm for resistance against anthracnose caused by *Colletotrichum lagenarium*. The Pharma Innovation Journal. 2021;**10**(4):404-406

[29] Parakkunnel R, Varalakshmi B, Sanna MK. Identification of sources of multiple disease resistance in advanced breeding lines of bottle gourd (*Lagenaria siceraria* (Molina) Standle). International Journal of Current and Applied Sciences. 2021;**10**(1):2511-2520

[30] AM AL-K, Kang B, Abbas A, Kaseb MO, Gu Q. Screening of resistance to cucumber green mottle mosaic virus in bottle gourd mutated by Ethyl Methane Sulphonate (EMS). Australasian Plant Pathology. 2022;**51**(5):535-541

[31] Bagri RK, Singh J, Goyal SK, Chawla N, Kumari M. Integrated disease management of downy mildew in bottle gourd (*Lagenaria siceraria* (Mol.) Stand). Journal of Entomology and Zoology Studies. 2019;**7**:315-317

[32] Bagri RK, Goyal SK, Singh J, Kumar V, Pryanka RS. Management of mosaic disease of Bottle Gourd (*Lagenaria siceraria* (Mol.) Stand) through integrated methods

[33] Bagri RK, Singh J, Goyal SK, Chawla N, Sharma RS, Sharma Y, et al. Management of *Cercospora* leaf spot in bottle gourd by integrated means under semi-arid conditions in Rajasthan. Journal of Pharmacognosy and Phytochemistry. 2019;**8**(3):4447-4449

[34] Tomar A, Mandloi S, Asati KP, Barche S. Integrated nutrient management in bottle gourd [*Lagenaria siceraria* (Mol) Standl.] variety kashi ganga on the plant disease incidence. TSS and Economics under Malwa Condition of Madhya Pradesh. International Journal of Current Science Research and Review. 2022;**5**(9):3298-3301

[35] Shahnaz E, Anwar A, Banday S. *Trichoderma* spp. as bio-stimulant: Molecular insights. In: New and Future Developments in Microbial Biotechnology and Bioengineering. Elsevier; 2022. pp. 337-350

[36] Rodrigues AO, May De Mio LL, Soccol CR. *Trichoderma* as a powerful fungal disease control agent for a more sustainable and healthy agriculture: Recent studies and molecular insights. Planta. Springer International Journal of Plant Biology. 2023;**257**(2):31

[37] Korate SR, Shete MH, Latake SB. In vitro and in vivo efficacy of fungicides and biocontrol agents in the control of fusarium moniliforme. The Incitant of Bottle Gourd Wilt. 2021:87-90

[38] Sultana N, Ghaffar A. Effect of fungicides, microbial antagonists and oil cakes in the control of *Fusarium oxysporum*, the cause of seed rot and root infection of bottle gourd and cucumber. Pakistan Journal of Botany. 2013;**45**(6):2149-2156

[39] Chaudhari KR, Patel NR, Patel AM. Eco-friendly management of *Fusarium* wilt of bottle gourd. The Pharma Innovation Journal. 2022;**SP11**(6):31-35

[40] Gaikwad SJ, Sen B, Meshram SU. Effect of bottlegourd seed coating with antagonists on seedlings, quantum of the pathogen inside the seedlings and population of the soil against *Fusarium oxysporum*. Plant and Soil. 1987;**101**:205-210

[41] Khanzada MA, Tanveer M, Maitlo SA, Hajano J, Ujjan AA, Syed RN, et al. Comparative efficacy of chemical fungicides, plant extracts and bio-control agents against *Fusarium solani* under laboratory conditions. Pakistan Journal of Phytopathology. 2016;**28**(2):133-139

[42] Joshi S, Borkar P, Joshi M. In vitro assessment of bio-agents, phyto extracts/organicals and fungicides against

Colletotrichum lagenarium. Trends in Biosciences;**12**(20):1321-1326

[43] Rajashree RP, Pawar R, Sapkal RT. Bioefficacy of antagonistic fungi *Trichoderma hammatum* and *Trichoderma harzianum* against *Alternaria alternata* (fr.) Keissler causing fruit rot in bottle gourd. Journal of Biological Control. 2014;**17**:234-236

[44] Chatur T, Anil G. Management of gummy stem blight of bottle gourd by bioagents. Journal of Mycology and Plant Pathology. 2014;**44**(2):225-226

[45] Patel MJ, Solanki VA, Rakholia KB, Khunt MD. Efficacy of biocontrol agents against *Didymella bryoniae* under field co. Trends in Biosciences. 2017;**10**(40):8547-8551

[46] Kumar S, Chugh RK, Singh J. Impact of spent mushroom substrate integration with *Trichoderma harzianum* on yield and gummosis disease incidence in bottle gourd (*Lagenaria siceraria*). International Journal of Plant & Soil Science. 2022;**34**(18):69-76

[47] Sultana N. Effect of fungicides and microbial antagonists in the control of *Lasiodiplodia theobromae*, the cause of seed rot, seedling and root infection of bottle gourd. Pakistan Journal of Agricultural Research. 2010;**23**(1-2):46-52

[48] Rahman MN, Tumpa FH, Islam AS, Khokon MA. Bio-priming of cucurbits and okra seeds with culture filtrates of *Trichoderma harzianum* for controlling seed-borne fungi. Journal of the Bangladesh Agricultural University. 2020;**18**(1):12-16

[49] Krishna V, Sm K. Development of seed coating formulation using consortium of *Bacillus subtilis* OTPB1 and *Trichoderma harzianum* OTPB3 for plant growth promotion and induction of systemic resistance in field and horticultural crops. Indian Phytopathology. 2015;**68**(1):25-31. Available from https://epubs.icar.org.in/index.php/IPPJ/article/view/46967

[50] Soni N, Raj K, Mor VS, Chugh RK. Mycoflora impact on germination and vigour of bottle gourd (*Lagenaria siceraria*) seeds and its management through different plant protectants. Seed Science and Technology. 2021;**49**(3):207-217

[51] Halder J, Sardana HR, Pandey MK, Nagendran K, Bhat MN. Synthesis and validation IPM technology and its economic analysis for bottle gourd (*Lagenaria siceraria*). Indian Journal of Agricultural Sciences. 2020;**90**(2):341-345

[52] Barua S, Molla AH, Haque MM, Alam MS. Performance of *Trichoderma*-enriched bio-organic fertilizer in N supplementation and bottle gourd production in field condition. Horticulture International Journal. 2018;**2**(3):106-114

[53] Mane PB, Mhase NL. Bioefficacy of different bioagents against root-knot nematode, *Meloidogyne incognita* infesting bottle gourd under laboratory conditions. International Journal of Plant Protection. 2017;**10**(1):87-91

[54] Faruk M. Development of bio-rational based management packages against root knot (*Meloidogyne incognita*) of bottle gourd. Bangladesh Journal of Agricultural Research. 2023;**46**(4):373-383

[55] Arie T, Namba S, Yamashita S, Doi Y, Kijima T. Biological control of fusarium wilt of bottle gourd by mix-cropping with welsh onion or chinese chive

inoculated with *Pseudomonas gladioli*, Japanese. Journal of Phytopathology. 1987;**53**(4):531-553

[56] Rani P, Singh M, Prashad H, Sharma M. Evaluation of bacterial formulations as potential biocontrol agents against the southern root-knot nematode, *Meloidogyne incognita*. Egyptian Journal of Biological Pest Control. 2022;**32**(1):29

[57] Tumpa FH, Sultana A, Alam MZ, Khokon MA. Bio-stimulation by seed priming with *Bacillus subtilis* for suppressing seed-borne fungal pathogens of vegetables in Bangladesh. Journal of the Bangladesh Agricultural University. 2016;**14**(2):177-184

[58] Lo CT. General mechanisms of action of microbial biocontrol agents. Plant Pathology Bulletin. 1998;7(4):155-166

[59] Manzar N, Kashyap AS, Goutam RS, Rajawat MV, Sharma PK, Sharma SK, et al. Trichoderma: Advent of versatile biocontrol agent, its secrets and insights into mechanism of biocontrol potential. Sustainability. 2022;**14**(19):12786

[60] Massart S, Jijakli HM. Use of molecular techniques to elucidate the mechanisms of action of fungal biocontrol agents: A review. Journal of Microbiological Methods. 2007;**69**(2):229-241

[61] Guetsky R, Shtienberg D, Elad Y, Fischer E, Dinoor A. Improving biological control by combining biocontrol agents each with several mechanisms of disease suppression. Phytopathology. 2002;**92**(9):976-985

[62] Xu XM, Jeffries P, Pautasso M, Jeger MJ. Combined use of biocontrol agents to manage plant diseases in theory and practice. Phytopathology. 2011;**101**(9):1024-1031

[63] Lutz MP, Wenger S, Maurhofer M, Défago G, Duffy B. Signaling between bacterial and fungal biocontrol agents in a strain mixture. FEMS Microbiology Ecology. 2004;**48**(3):447-455

[64] Kotasthane A, Agrawal T, Kushwah R, Rahatkar OV. In-vitro antagonism of *Trichoderma* spp. against *Sclerotium rolfsii* and *Rhizoctonia solan*i and their response towards growth of cucumber, bottle gourd and bitter gourd. European Journal of Plant Pathology. 2015;**141**:523-543

[65] Zhang F, Liu C, Wang Y, Dou K, Chen F, Pang L, et al. Biological characteristic and biocontrol mechanism of *Trichoderma harzianum* T-A66 against bitter gourd wilt caused by *Fusarium oxysporum*. Journal of Plant Pathology. 2020;**102**:1107-1120

[66] Na SM, Abid M, Rehman S, Ashraf M, Zhang L, Anees M. Chitinolytic activity of the indigenous *Trichoderma* spp. from the north west of Pakistan against the fungal phytopathogens. Pakistan Journal of Botany. 2019;**51**(2):711-716

[67] Shah P, Abid M, Abbas A, Basheer N, Sheikh AH, Jabeen N, et al. Screening of *Trichoderma viride* as a bio-control agent against different soil-borne infectious and phytopathogenic fungi. International Journal of Biology and Biotechnology. 2020;**17**(2):405-410

[68] Laura G-G, Zeriouh H, Romero D, Cubero J, de Vicente A, Pérez-García A. The antagonistic strain B acillus subtilis UMAF 6639 also confers protection to melon plants against cucurbit powdery mildew by activation of jasmonate- and salicylic acid-dependent defence responses. Microbial Biotechnology. 2013;**6**(3):264-274

[69] Romero D, De Vicente A, Zeriouh H, Cazorla FM, Fernández-Ortuño D,

Torés JA, et al. Evaluation of biological control agents for managing cucurbit powdery mildew on greenhouse-grown melon. Plant Pathology. 2007;**56**(6):976-986

[70] Romero D, De Vicente A, Olmos JL, Dávila JC, Pérez-García A. Effect of lipopeptides of antagonistic strains of Bacillus subtilis on the morphology and ultrastructure of the cucurbit fungal pathogen Podosphaera fusca. Journal of Applied Microbiology. 2007;**103**(4):969-976

[71] Gafni A, Calderon CE, Harris R, Buxdorf K, Dafa-Berger A, Zeilinger-Reichert E, et al. Biological control of the cucurbit powdery mildew pathogen Podosphaera xanthii by means of the epiphytic fungus *Pseudozyma aphidis* and parasitism as a mode of action. Frontiers in Plant Science. 2015;**6**:132

[72] Ongena MA, Daayf F, Jacques P, Thonart P, Benhamou N, Paulitz TC, et al. Protection of cucumber against Pythium root rot by fluorescent pseudomonads: predominant role of induced resistance over siderophores and antibiosis. Plant Pathology. 1999;**48**(1):66-76

[73] Zeriouh H, Romero D, García-Gutiérrez L, Cazorla FM, de Vicente A, Pérez-García A. The Iturin-like lipopeptides are essential components in the biological control arsenal of Bacillus subtilis against bacterial diseases of cucurbits. Molecular Plant-Microbe Interactions. 2011;**24**(12):1540-1552

[74] Romero D, De Vicente A, Rakotoaly RH, Dufour SE, Veening JW, Arrebola E, et al. The Iturin and fengycin families of lipopeptides are key factors in antagonism of Bacillus subtilis toward Podosphaera fusca. Molecular Plant-Microbe Interactions. 2007;**20**(4):430-440

Section 3

Abiotic Stress in *Cucurbitaceae* Crops

Chapter 5

Fruit Cracking in Melon

Lija Maryomana and Salmabeevi Suhara Beevy

Abstract

Melon (*Cucumis melo* L.) is an economically profitable crop in temperate and tropical regions. Melons vary in shape, size, and skin texture; they are classified under different varieties. Fruit cracking is a general problem of melon fruit grown worldwide. This physiological disorder intensively affects the production and marketable value of fruits. Studies revealed that fruit cracking causes a 70% loss in the economy of the melon fruit industry. The cracking becomes more visible when the fruits reach maturity; no single factor is known to prevent it effectively. The severity of fruit cracking depends on the nature of the variety, climate zone, where the variety is grown, fruit growing patterns, and cultural practices. It has also been linked to improper irrigation, environmental factors, and nutritional deficiencies, particularly boron, calcium, zinc, and potassium. Horticultural practices, such as spraying growth promoters, micronutrients, antitranspirant, and regular drip irrigation with mulching, have been recommended to avoid fruit cracking in melon. Although fruit cracking is a significant economic risk, research on its cause and management in melon is limited compared to other fruit crops. Hence, the present chapter summarizes the underlying causes of melon fruit cracking and potential control strategies to reduce melon fruit cracking.

Keywords: *Cucumis melo*, fruit cracking, physiological disorder, nutrient deficiency, morphological characters

1. Introduction

Melon (*Cucumis melo* L.), a major cucurbit cultivated around the world, exhibits significant morphological, physiological, and molecular variations at the cultivar subgroup level [1–3]. Considering the most significant character, i.e., ovary harness, that distinguishes the taxa, the species *Cucumis melo* is divided into subspecies *Cucumis melo* ssp. *melo* and *Cucumis melo* ssp. *agrestis* at the intraspecific level [4–6]. This subspecies was further classified into different botanical groups or variants, resulting in quadrinomial nomenclature (e.g., *Cucumis melo* ssp. *agrestis* var. *momordica*) [7] as per the studies done by Hammer et al.; Pitrat et al., Pandey and Anju [8–10] five melon types are included under *Cucumis melo* ssp. *Agrestis,* while 12 botanical variants are enlisted in *Cucumis melo* ssp *melo*. The intraspecific classification of melon varieties is depicted in **Figure 1** [11].

Melon cracking is a physiopathy that causes both internal and external changes in the fruit by disrupting the water balance and nutrient homeostasis. Fruit cracking occurs naturally at the end of fruit development, just after ripening and before seed dispersal [12]. It is a physiological disorder that affects the exocarp and mesocarp and can be distinguished from epidermis cracking, which is more superficial and includes the

Cucumis melo Sub.sp melo (appresed, very short ovary hairs)	*Cucumis melo Sub.sp agrestis* (Long spreading, ovary hairs)
1. *C.melo* var.*cantalupensis* Naudin	1. *C.melo* var.*momordica* Roxb.
2.*C.melo* var.*reticulatus* Ser	2.*C.melo*var.*acidulus*Naudin
3. *C.melo* var.*flexuosus* L.	3. *C.melo* var.*conomon* Thunb
4. *C.melo* var.*inodorus* H.Jac.	4. *C.melo* var.*makuwa* Makino
5. *C.melo* var.*chandalak* Gabaev	5. *C.melo* var.*chinensis* Pangalo
6. *C.melo* var.*adana* Pangalo	
7. *C.melo* var.*ameri* Pangalo	
8. *C.melo* var.*chate* Hasselq.	
9. *C.melo* var.*chito* Morren	
10. *C.melo* var.*dudaim* L.	
11. *C.melo* var. *tibish* Mohamed	

Figure 1.
Intraspecific classification of Cucumis melo.

cuticle and epidermal tissue [13]. Deeper cracking is distinguished by an opening to the pulp's interior, known as splitting [14, 15]. The fruit rind is pertinent for crack resistance, portability, storability, and quality during storage (shelf-life quality) [16]. Thin-rind varieties, such as var. *momordica*, var. *conomon*, and var. *hami*, are more prone to cracking and causing more damage during transportation or retail display. As part of the dehiscent mechanism, climacteric varieties are predisposed to blossom end cracking [17].

Fruit cracking may be caused by high evapotranspiration, low relative air humidity (R.H.), water imbalance, and sharp temperature fluctuations during fruit growth and development [18]. It has also been linked to improper irrigation, environmental factors, and nutritional deficiencies, particularly boron, calcium, zinc, and potash [19, 20]. When the fruits reach maturity, the cracking becomes more visible [21, 22]. Several horticultural practices, such as spraying growth promoters, micronutrients, antitranspirants, and regular drip irrigation and mulching, have been recommended for fruit cracking management [23].

This book chapter reviews the research and information on fruit cracking, an essential physiological disorder of melon varieties, in terms of different patterns of cracking, factors associated with it, and management practices to control its intensity.

Figure 2.
(A) Pattern of fruit cracking in Indian snapmelon. (B) Fruit cracking in culinary melon and water melon.

2. Patterns of melon fruit cracking

Cracking is a notable disorder that can result in significant losses of marketable yield and revenue in the fruit industry. Variations can also be seen in the cracking pattern. Longitudinal cracking, burst cracking, ring or concentric cracking, crazing or russeting, star or radial cracking, and core failure are the different types of fruit cracking in melon. Snap melon has a wide variety of cracking patterns (**Figure 2a** and **b**) [24]. Melon fruit cracking or splitting around the distal (blossom-end) is thought to be a seed dispersal mechanism associated with post-climacteric senescence [17, 25–27]. These include both longitudinal and random cracking patterns that begin around the fruit's equatorial region.

3. Factors affecting fruit cracking in melon

The severity of melon fruit cracking depends on the nature of the variety, the climate zone where the variety is grown, fruit growing patterns, cultural practices, etc. It has also been linked to improper irrigation, environmental factors, and nutritional deficiencies, particularly boron, calcium, zinc, and potassium (**Figure 3**).

3.1 Plant factors

3.1.1 Plant morphological characteristics

Fruit cracking is influenced by the morphology of the fruit's surface. Characteristics such as netting, netting density, and sutures all play essential roles in melon fruit cracking. Deeply sutured melon cultivars are more vulnerable than non-sutured melon cultivars. The deep, sutured, and thin rind of Canary melon causes it to crack early. Thick cuticle deposition on the netted region reduces rind elasticity during fruit expansion, whereas rapid water intake raises turgor pressure in the flesh and makes the fruit more susceptible [28].

The shape of melon fruits impacts cracking. Martinez et al. [17] reported that round melons are less susceptible to cracking than oblong melons. Many morphological

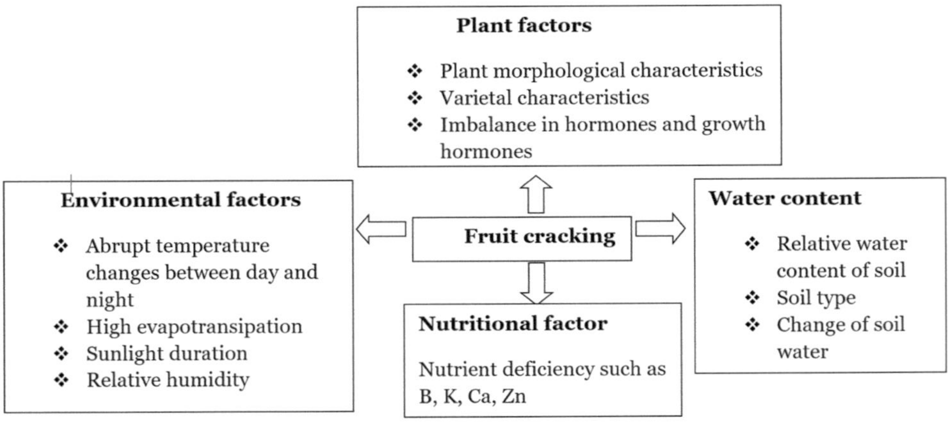

Figure 3.
Factors affecting fruit cracking.

characteristics influence cracking sensitivity, including cuticle thickness and physical properties, the number of hypodermal layers [29], fruit shape, and fruit size [30–32]. However, calcium content and pectin value, cell wall structure and component [33], and the quantity and volume of intercellular spaces could all influence these properties [34]. Skin's biomechanical properties play an important role in maintaining internal pressure and resisting fruit cracking. According to Lane et al. [35], the fruit skin's role in cracked resistance is related to the calcium content of epidermis cells, which increases cell integrity.

Many physical characteristics contribute to fruit cracking resistance. The sensitive cultivar had thinner skin, a higher seed-to-skin weight ratio, and was significantly smaller in size. According to Aloni et al. [36] & Matthews et al. [37], when exposed to water stress, peel mechanical properties change, with decreased peel extensibility leading to thicker and stiffer peel. These changes may account for the cracking of pomegranate fruits [38].

3.1.2 Varietal influence on melon fruit cracking

Fruit cracking susceptibility is thought to be inherited and the severity of cracking varies greatly between varieties; Cuartero et al. [39] discovered that genetic characteristics influence fruit cracking. Fruit cracking is more or less common in different cultivars [40]. Multiple genes regulate fruit cracking [41]. Intraspecific classification of melon species revealed about 16 varieties of melon, with significant variation in their cracking pattern. Thin rind melon types, such as *Hami* melon, *snapmelon,* and *conomon* group, were found to be more prone to cracking than thick rind melon types such as *reticulatous indorous* and group. The high intensity of cracking may also be due to Hami melon's high sweetness (total soluble solids 11–15%). Cracking disorder in thin-rind cultivars is exacerbated when cracks form after harvest, during transportation, or during retail display. As a result, such cultivars require special handling during transportation, in the laboratory and at retail.

Snapmelon, an Indian native, is also known as 'Phoot,' which means 'to split,' and has a variety of cracking patterns. In this variety, the presence of incipient fruit cracking or splitting indicates harvest readiness or fruit maturity [42]. Cracking occurs in the *Charentais*-type cultivar *Vedrantais* as a result of high ethylene production as part of the seed dispersal mechanism [25]. During harvest or transportation, an explosive type of cracking was observed in an indorous variety, known as Piel de Sapo-type cultivar T111. It is a cultivar with deep sutures that cracks easily [43].

3.1.3 Hormonal influence on melon fruit cracking

Ethylene is a naturally occurring plant growth regulator that has a variety of effects on the growth, development, and storage life of many fruits. Climacteric fruit ripening in melon is associated with induced ethylene biosynthesis. Ethylene peak within 2 hours in the senescent overripe fruits of *snapmelon* and *cantaloupe* melon fruit in the open field is usually associated with stem end cracking in. According to Fernandez Trujillo [44], a small burst of ethylene production more significant than 1–4 Pmol/kg/g in melon is a sign of fruit splitting or other mechanical damage.

Fruit cracking has also been linked to an imbalance of auxins, gibberellins, and cytokinins in various crops [45]. Normal fruits contain more gibberellins and less ABA [41, 42], and an imbalance between the two causes fruit cracking [46]. Gibberellin (GA3) can increase cell wall plasticity, promoting cell growth and cell extension and possibly preventing fruit creasing [47]. Yilmaz and Ozguven [48] discovered that the ABA content of the peel was higher in cracked fruits than in healthy (non-cracked) fruit.

3.2 Environmental factor

According to Ikram et al. [49], direct sunlight raises the temperature and evapotranspiration of the fruit surface, resulting in high moisture loss and increased cracking susceptibility. Fruit cracking susceptibility increases in the field when the day temperature exceeds the light temperature at the end of the ripening process [50]. Intense solar radiation significantly raises the temperature of the fruit, which raises the internal turgor pressure of the pulp inside the rind. Landg and During [51] suggested the effect of intense solar radiation as a cause of melon fruit cracking.

A high relative humidity (R.H.) prevents transpiration, which causes a high pressure inside the fruit, causing the epidermis to crack [52], especially during prolonged periods of 99–100% R.H., alone or in combination with rain. Due to transpiration-inhibiting effects, high relative humidity conditions in the air have been observed to cause fruit cracking, both in the field and in storage [15]. This effect becomes more pronounced whenever nighttime temperatures drop [53]. Fruit cracking can occur during storage due to abrupt changes in R.H. or temperature [15], mainly when the R.H. is high.

Temperature increases, hot, dry winds, a downpour after a dry season, and significant differences in day-night temperatures with temperatures greater than 38°C combined with an R.H. of 60% all favor fruit cracking in most of the crops [54].

3.3 Relative water content on the fruit surface and soil

Fruit cracking is caused primarily by high water content in fruit, especially during the monsoon season or after the first rains after a dry season [15]. Excessive water can cause classic watermelon fruit split incidents. This is common in melons, as well as tree fruits such as plums. Irrigation must be applied evenly, especially during the last 2 weeks of growth. Too much water at this stage may cause the fruit to crack.

Fruit cracking in apples and sweet cherries occurs as a result of excessive cell enlargement of the fruits following a significant increase in soil moisture [55]. The strength of the skin is affected by changes in soil moisture during fruit development. When the moisture content of the soil is reduced, skin strength improves. Skin strength, on the other hand, decreases with increasing soil moisture content [56, 57]. The plant water gradient induces the process of water loss from the plant to the atmosphere *via* the fruit surface or leaf area surface in the soil-plant-atmosphere continuum, which dramatically influences water relations, and thus leads to the incidence of cracking in fruits because the high demand for water during fruit growth and development is met by the leaf *via* the source-sink relationship [58]. The development of high hydrostatic pressure in the fruit (turgor pressure) more remarkable than the tensile strength of the cell walls under conditions of high-water availability and low evaporative demand has been attributed to the cracking or splitting of fruits caused by unseasonal rainfall [59].

3.4 Nutritional factors

Fruit cracking is caused by excessive nitrogen fertilization combined with insufficient potassium or calcium in the field [60]. Premature cracking in melon cultivars is caused by an insufficient supply of nitrogen and potassium in the open field (**Figure 4**) [39]. The phosphorus (P) content of peel during the fruit cracking period is positively related to the rate of fruit cracking. The fruit peel becomes thinner as the P level rises. When citrus fruit lacks phosphorous, it shrinks, and the peel becomes rougher.

Figure 4.
Premature cracking in var. momordica and var. acidulus.

Potassium can maintain high osmotic and turgor pressures, which can provide energy for cell division, cell wall extension, and cell expansion, thereby accelerating cell growth. A high potassium content can increase fruit size and make the peel thick and smooth; on the other hand, a low potassium content can cause fruit cracking and dropping, resulting in smaller fruit, thinner peel, and a reduction in soluble solids content (SSC), organic acids, and vitamin C [61]. However, some studies have found that the potassium content of cracked fruit peels is higher than that of regular fruit peels [62].

Boron is essential for the promotion of cell division and the synthesis of cell walls [63]. Because of the association of cell wall composition, boron may aid in the maintenance of cell wall integrity and toughness [64–66]. According to Wu et al. [67], boron can enhance the regulatory function of endogenous growth regulators. Increased boron content in the fruit peel reduces fruit cracking.

Calcium is required for proper plant growth and development because it performs metabolic functions in nutrient uptake, as well as abiotic and biotic stress resistance. Calcium, an important component of cell walls, contributes to cell wall cohesiveness and strength [64], and its concentration was higher in normal fruit than in cracked fruit [68]. Calcium is most likely involved in the increased elasticity, strength, and thickness of epidermal cell walls. Calcium also aids in the deposition of pectin, making fruit more resistant to cracking under the higher rates of turgor pressure seen during water stress. Calcium and magnesium are in charge of strengthening the bonds between epidermal and other fruit cells, resulting in increased strength and decreased cracking [69]. Furthermore, Ca's role in preventing the formation of an abscission zone between fruit pedicles and bearing branches, as well as regulating enzyme activity and photosynthesis [70–72], could result in fruit splitting percentage control.

Zinc is also essential in regulating water absorption by plant roots. Fruit cracking has been reduced by a concurrent antitranspirant. Antitranspirants reduce transpiration and regulate fruit skin elasticity by reducing stomatal opening and increasing leaf resistance to water vapor diffusion without affecting carbon dioxide uptake. It also reduces the absorption of radiant energy, lowering leaf temperatures, and transpiration rates; wax emulsions on fruit form thin transparent films that hinder the escape of water vapor from the fruit surface and prevent stomata from opening fully and lowering transpiration [73]. These may be possible reasons for lowering the incidence of fruit cracking in pomegranates.

4. Managemental practices to control fruit cracking

- Orchard floor management should be implemented to prevent rapid changes in soil moisture to protect susceptible fruits from cracking [74]. Drip irrigation, mulching with organic and inorganic sources, spreading compost or manure over the soil, green manuring, and growing cover crops are all good ways to conserve soil moisture by reducing evaporation, protecting the soil from heat and sunlight, and making good use of the available water.

- Cracking occurs due to excess solar radiation, and sunburn can be minimized by shade nets, bagging of individual fruits, or spraying the entire crop with particle films such as kaolin, aluminum silicate clay, or calcium carbonate [75]. These shade nets and particle film reduces radiation stress in crops by protecting the foliage and fruit from ultraviolet and infrared radiation. Excessive evapotranspiration from the fruit surface results in excessive moisture loss when exposed to direct heat. Bagging also aids in the prevention of cracking, the protection of fruit from insects or pests, and the improvement of fruit quality [76].

- Growth regulators, such as forchlorfenuron (CPPU), a cytokinin, and benzothiadiazole, are effectively used in watermelon and Hami melon, respectively, in China in order to control premature cracking.

- Balanced nutrition: Adequate potassium and calcium can promote the development of thicker rinds that are less susceptible to cracking because potassium and calcium play a crucial role in membrane and cell wall integrity. Foliar application of calcium and potassium during fruit development and the maturation stage improves melon fruit texture through turgor maintenance [77].

- Foliar application of micronutrients, such as calcium, zinc, and boron, is effectively used to control melon craking in *C. melo* cv. Grand Riado (Sakata) in Spain [78].

- Recent advances in nanotechnology indicate that using nanomaterials in agriculture may aid in increasing productivity. Furthermore, nanofertilizers and nanonutrients can help to reduce fruit cracking. Davarpanah et al. [79] investigated the effects of foliar fertilization with nano-nitrogen and urea fertilizers containing nanoparticles (nN) on pomegranates, finding that fruit quality and yields improved. Furthermore, they also reported that foliar treatment with the nano-calcium fertilizer significantly reduced fruit cracking in pomegranates when compared to the control treatment.

- Grafting of susceptible cultivars with resistant rootstock with a vigor root system reduces the incidence of cracking [80].

5. Conclusion

The problem of fruit cracking in melons is a complex phenomenon that is caused by a variety of factors. Advances in research on fruit cracking in terms of causes, physiology, and effective management clearly demonstrate that the problem cannot be considered in isolation but that both external and internal conditions must be considered holistically. For fruit cracking management, horticultural practices such as spraying growth promoters, micronutrients, antitranspirants, and regular drip irrigation with mulching have been recommended. Each factor can only be advocated as being sufficiently effective in controlling fruit cracking. At best, the role of integrated orchard management, which aims to reduce water, nutrition, and physiological factors that contribute to fruit cracking, should be considered. Several studies have confirmed that nanoparticle-based fertilizers have the potential to increase crop yield and quality under various biotic and abiotic stress conditions. However, the use of nano-based fertilizers to prevent fruit cracking is still in its infancy. Despite the fact that it is a natural occurrence, proper monitoring and development of accurate management practices are required to overcome the severity of this natural threat and avoid massive economic loss in the melon fruit industry.

Author details

Lija Maryomana* and Salmabeevi Suhara Beevy
Department of Botany, University of Kerala, Thiruvananthapuram, Kerala, India

*Address all correspondence to: lijaomana@gmail.com

References

[1] Stepansky A, Kovalski I, Perl-Treves R. Intraspecific classification of melons (*Cucumis melo* L.) in view of their phenotypic and molecular variation. Plant Systematics and Evolution. 1999;**217**(3):313-322. DOI: 10.1007/BF00984373

[2] Naudin C. Essays on a monograph of species and varieties of the genus Cucumis. Annales des Sciences Naturelles. 1859;**11**:5-87

[3] Munger HM, Robinson RW. Nomenclature of *Cucumis melo* L. Cucurbit Genetics Cooperative Report. 1991;**14**:43-44

[4] Kirkbride JH. Biosystematic Monograph of the Genus Cucumis (Cucurbitaceae): Botanical Identification of Cucumbers and Melons. Parkway Publishers, Inc.; 1993

[5] Jeffrey C. A review of the Cucurbitaceae. Botanical Journal of the Linnean Society. 1980;**81**(3):233-247. DOI: 10.1111/j.1095-8339.1980.tb01676.x

[6] Jeffrey C. Cucurbitaceae. In: Hanelt P, Institute of Plant Genetics and Crop Plant Research, editors. Mansfeld's Encyclopedia of Agricultural and Horticultural Crops. Vol. 3. Berlin: Springer-Verlag; 2001. pp. 1510-1557

[7] Pitrat M, Hanelt P, Hammer K. Some comments on infraspecific classification of cultivars of melon. In: VII Eucarpia Meeting on Cucurbit Genetics and Breeding. Vol. 510. 2000. pp. 29-36. DOI: 10.17660/ActaHortic.2000.510.4

[8] Hammer K, Gladis T. Notes on infraspecific nomenclature and classifications of cultivated plants in Compositae, Cruciferae, Cucurbitaceae, Gramineae (with a remark on Triticum dicoccon Schrank) and Leguminosae. Genetic Resources and Crop Evolution. 2014;**61**(8):1455-1467. DOI: 10.1007/s10722-014-0148-8

[9] Pitrat M. Melon genetic resources: Phenotypic diversity and horticultural taxonomy. In: Genetics and Genomics of Cucurbitaceae. Cham: Springer; 2016. pp. 25-60. DOI: 10.1007/7397_2016_10

[10] Pandey A, Rajkumar S. A new potential variety of cultivated melon (*Cucumis melo* L.) from north western India. Genetic Resources and Crop Evolution. 2021;**68**(2):785-794. DOI: 10.1007/s10722-020-00997-2

[11] Lija M, Beevy SS. A review on the diversity of melon. Plant Science Today. 2021;**8**(4):995-1003

[12] Lichter A, Dvir O, Fallik E, Cohen S, Golan R, Shemer Z, et al. Cracking of cherry tomatoes in solution. Postharvest Biology and Technology. 2002;**26**(3):305-312

[13] Yu J, Zhu M, Bai M, Xu Y, Fan S, Yang G. Effect of calcium on relieving berry cracking in grape (*Vitis vinifera* L.)'Xiangfei'. PeerJ. 2020;**8**:e9896

[14] Opara LU, Studman CJ, Banks NH, Opara UL. Fruit skin splitting and cracking. Horticultural Reviews. 2010;**19**:217-262

[15] Cooman A, Torres C, Fischer G. Determining the cause of cape gooseberry (*Physalis peruviana* L.) fruit cracking in greenhouse conditions. II. Effects of calcium, boron and copper supply. Agronomía Colombiana. 2005;**23**(1):74-82

[16] Wang Y, Guo L, Zhao X, Zhao Y, Hao Z, Luo H, et al. Advances

in mechanisms and omics pertaining to fruit cracking in horticultural plants. Agronomy. 2021;**11**(6):1045

[17] Martínez JA, Jowkar MM, Obando-Ulloa JM, Varó P, Moreno E, Monforte AJ, et al. Uncommon disorders and decay in near-isogenic lines of melon and reference cultivars. Horticultura Brasileira. 2009;**27**:505-514

[18] Abd IE, Rahman EI. Physiological studies on cracking phenomena of pomegranate. Journal of Applied Sciences Research. 2010;**6**:669-703

[19] Gharesheikhbayat R. Anatomical study of fruit cracking. In: Pomegranate CV. Malas-e-Torsh. Pajouhesh Va Sazandegi J; 2006;**18**:10-14

[20] Saei H, Sharifani MM, Dehghani A, Seifi E, Akbarpour V. Description of biomechanical forces and physiological parameters of fruit cracking in pomegranate. Scientia Horticulturae. 2014;**178**:224-230

[21] Hoda AK, Hoda SHA. Cracking and fruit quality of pomegranate (*Punica granatum* L.) as affected by pre-harvest sprays of some growth regulators and mineral nutrients. Journal of Horticultural Science & Ornamental Plants. 2013;**5**:71-76

[22] Yazici K, Kaynak L. Effects of kaolin and shading treatments on sunburn on fruit of Hicaznar cultivar of pomegranate (Punica granatum L. cv. Hicaznar). In: International Symposium on Pomegranate and Minor Mediterranean Fruits. 16 Oct 2006;**818**:167-174

[23] Waskar DP. Pomegranate (*Punica granatum* L.). In: Saroj PL, Awasthi OP, editors. Advances in arid Horticulture. Lucknow: International Book Distributing Company; 2006. pp. 375-394

[24] Singh A, Shukla AK, Meghwal PR. Fruit cracking in pomegranate: Extent, cause, and management – A review. International Journal of Fruit Science. 2020;**20**(sup3):S1234-S1253

[25] Dhillon NPS, Singh J, Fergany M, Monforte AJ, Sureja AK. Phenotypic and molecular diversity among landraces of snapmelon (*Cucumis melo* var. *momordica*) adapted to the hot and humid tropics of eastern India. Plant Genetic Resources. 2009;**7**(3):291-300

[26] Corbacho J, Romojaro F, Pech JC, Latché A, Gomez-Jimenez MC. Transcriptomic events involved in melon mature-fruit abscission comprise the sequential induction of cell-wall degrading genes coupled to a stimulation of endo and exocytosis. PLoS One. 2013;**8**(3):e58363

[27] Fernández-Trujillo JP, Obando J. Mapping fruit susceptibility to postharvest physiological disorders and decay using a collection of near-isogenic lines of melon. Journal of the American Society for Horticultural Science. 2007;**132**:739-748

[28] Vrebalov J, Pan IL, Arroyo AJM, McQuinn R, Chung M, Poole M, et al. Fleshy fruit expansion and ripening are regulated by the tomato SHATTERPROOFgene TAGL1. Plant Cell. 2009;**21**:3041-3062

[29] Keren-Keiserman A, Tanami Z, Shoseyov O, Ginzberg I. Differing rind characteristics of developing fruits of smooth and netted melons (*Cucumis melo*). The Journal of Horticultural Science and Biotechnology. 2004;**79**:107-113

[30] Bargel H, Neinhuis C. Tomato (*Lycopersicon esculentum* Mill.) fruit growth and ripening as related to the biomechanical properties of fruit skin and isolated cuticle. Journal of Experimental Botany. 2005;**56**(413):1049-1060

[31] Emmons CLW, Scott JW. Ultrastructural and anatomical factors associated with resistance to cuticle cracking in tomato (*Lycopersicon esculentum* Mill.). International Journal of Plant Sciences. 1998;**159**(1):14-22

[32] Koske TJ, Pallas JE, Jones JB. Influence of ground bed heating and cultivar on tomato fruit cracking 1. HortScience. 1980;**15**(6):760-762

[33] Sekse L. Fruit cracking in Norwegian grown sweet cherries. Acta Agriculturae Scandinavica. 1987;**37**(3):325-328

[34] Cybulska J, Zdunek A, Konstankiewicz K. Calcium effect on mechanical properties of model cell walls and apple tissue. Journal of Food Engineering. 2011;**102**:217-223. DOI: 10.1016/j.jfoodeng.2010.08.019

[35] Chanliaud E, Burrows KM, Jeronimidis G, Gidley MJ. Mechanical properties of primary plant cell wall analogues. Planta. 2002;**215**:989-996. DOI: 10.1007/s00425-002-0783-8

[36] Lane WD, Meheriuk M, McKenzie DL. Fruit cracking of a susceptible, an intermediate, and a resistant sweet cherry cultivar. HortScience. 2000;**35**(2):239-242

[37] Aloni B, Karni L, Moreshet S, Yao C, Stanghellini C. Cuticular cracking in bell pepper fruit: II. Effects of fruit water relations and fruit expansion. The Journal of Horticultural Science and Biotechnology. 1999;**74**(1):1-5

[38] Matthews MA, Cheng G, Weinbaum SA. Changes in water potential and dermal extensibility during grape berry development. Journal of the American Society for Horticultural Science. 1 Mar 1987;**112**(2):314-319

[39] Lawlor DW, Leach JE. Leaf growth and water deficits: Biochemistry in relation to biophysics. In: Baker NR, Davies WJ, Ong CK, editors. Control of Leaf Growth. Soc. Expt. Biol. Vol. 5. Society for Experimental Biology; 1985. pp. 267-294

[40] Cuartero J, Palomares G, Balasch S, Nuez F. Tomato fruit cracking under plastic-house and in the open air. II. Generaland specific combining abilities. In: Genetics and Breeding of Tomato, Proceedings of the Meeting of the Eucarpia Tomato Working Group, Avignon, France, 18-21 May 1981. Versailles, France: Institut National de la Recherche Agronomique; 1981

[41] Beyer M, Hahn R, Peschel S, Harz M, Knoche M. Analysing fruit shape in sweet cherry *Prunus avium* L. Scientia Horticulturae. 2002;**96**:139-150. DOI: 10.1016/S0304-42380200123-1

[42] Vaidyanathan S, Harrigan GG, Goodacre R. Metabolome Analyses, Strategies for Systems Biology. Berlin, Germany: Springer Science and BusinessMedia; 2006

[43] Flores F, Martı'nez-Madrid MC, Sa'nchez-Hidalgo FJ, Romojaro F. Differential rind and pulp ripeningof transgenic antisense ACC oxidasemelons. Plant Physiology and Biochemistry. 2001;**39**:37-43

[44] Fernández-Trujillo JP, Lester GE, Dos-Santos N, Martínez JA, Esteva J, Jifon JL, et al. Pre-and postharvest muskmelon fruit cracking: Causes and potential remedies. HortTechnology. 2013;**23**(3):266-275

[45] Rai M, Dey P, Nath V, Das B. Litchi Production Technology. Ranchi: Horti. and Agroforestry Res. Prog; 2002. p. 21

[46] Sharma KK, Dhillon WS. Evaluation of evergreen varieties of pomegranate under Punjab conditions. Agricultural Science Digest. 2002;**22**:42-44

[47] Holtzhausen LC. Creasing: Formulating a hypothesis. Proceedings of International Society of Citriculture. 1981;**1**:201-204

[48] Yilmaz C, Ozguven A. Hormone physiology of preharvest fruit cracking in pomegranate (*Punica granatum*). Acta Horticulturae. 2003;**727**:135-139. [Google Scholar]

[49] Ikram S, Jaskani MJ, Ikram S, Qureshi MA, Rehman SU, Hussain M, et al. Effectiveness of pollen germination media for improving storage potential in pomegranate (*Punica granatum*) Germplasm. Journal of Plant and Environment. 30 Dec 2022;**4**(2):155-161

[50] Yılmaz C, Özgüven AI. The effects of some plant nutrients, gibberellic acid and pinolene treatments on the yield, fruit quality and cracking in pomegranate. International Symposium on Pomegranate and Minor Mediterranean Fruits. 2006;**818**:205-212

[51] Lang A, During H. Grape berry splitting and some mechanical properties of the skin. Vitis. 1990;**29**:61-70

[52] Opara LU, Studman CJ, Banks NH. Sunlight affects the incidence of internal ring cracking and other physical attributes of 'Gala'apples. Journal of Tree Fruit Production. 1997;**2**(1):45-52

[53] Fischer G, Ramírez F, Casierra-Posada F. Ecophysiological aspects of fruit crops in the era of climate change. A review. Agronomía Colombiana. 2016;**34**(2):190-193

[54] Elfving DC, Kaufmann MR. Diurnal and seasonal effects of environmental on plant water relations and fruit diameter of citrus. Journal of the American Society for Horticultural Science. 1972;**118**:562-570

[55] Gourley JB, Howlett FS. Modern Fruit Production. N.Y.: The Mac Millan Coy; 1941. pp. 443-444

[56] Kamimura SH. A method of testing crack resistance in tomatoes. Japan Agricultural Research Quarterly. 1977;**11**(2):111-114

[57] Peet MM. Fruit cracking in tomato. HortTechnology. 1992;**2**(2):216-223

[58] Hepaksoy S, Aksoy U, Can HZ, Ui MA. Determination of relationship between fruit cracking and some physiological responses, leaf characteristics and nutritional status of some pomegranate varieties. Options Méditerr. Sér. A. 2000;**42**:87-92

[59] Considine J, Brown K. Physical aspects of fruit growth: Theoretical analysis of distribution of surface growth forces in fruit in relation to cracking and splitting. Plant Physiology. 1981;**68**(2):371-376

[60] Cabello MJ, Castellanos MT, Tarquis AM, Cartagena MC, Arce A, Ribas F. Determination of the uptake and translocation of nitrogen applied at different growth stages of a melon crop (*Cucumis melo* L.) using N-15 isotope. Scientia Horticulturae. 2011;**130**:541-550

[61] Alva AK, Mattos D Jr, Paramasivam S, Patil B, Dou H, Sajwan KS. Potassium management for optimizing citrus production and quality. International Journal of Fruit Science. 2006;**6**(1):3-43

[62] Storey JB. Research notes. Pecan Quarterly. 1969;**3**:20

[63] Juan LI, Jiezhong CH. Citrus fruit-cracking: Causes and occurrence. Horticultural Plant Journal. 2017;**3**(6):255-260

[64] Li J, Zhang PP, Chen JZ, Yao Q, Jiang YM. Cellular Wall metabolism in citrus fruit pericarp and its relation to creasing fruit rate. Scientia Horticulturae. 2009;**122**:45-50

[65] Goldbach HE, Wimmer MA. Boron in plants and animals: Is there a role beyond cell-wall structure. Journal of Plant Nutrition and Soil Science. 2007;**170**:39-48

[66] Dong Y, Yang X, Liu J, Wang BH, Liu BL, Wang YZ. Pod shattering resistance associated with domestication is mediated by a NAC gene in soybean. 2009. Nature Communications. 2014;**5**:3352

[67] Wu ZR, Chen JC, Chen WS. 'Duwei' pomelo fruit cracking and its control. China Citrus. 1987;**2**:23

[68] Brett C, Waldron K. Physiology and Biochemistry of Plant Cell Walls. London: Unwin Hyman; 1990

[69] Huang XM, Wang HC, Lu XJ, Yuan WQ, Lu JM, Li JG, et al. Cell wall modifitions in the pericarp of litchi (*Litchi chinensis* Sonn.) cultivars that differ in their resistance to cracking. The Journal of Horticultural Science and Biotechnology. 2006;**81**:231-237

[70] Li JG, Huang HB. Physico-chemical properties and peel morphology in relation to fruit-cracking susceptibility in litchi fruit. Journal of South China Agricultural University. 1995;**16**(1):84-89

[71] Sanyal SJR, Hasan A, Ghosh B, Mitra SK. Studies on sun-browning and skin cracking in some varieties of litchi. Indian Agriculturist. 1990;**34**(1):19-23

[72] Choi JH, Lee B, Gu M, Lee UY, Kim MS, Jung SK, et al. Course of fruit cracking in 'Whansan' pears. Horticulture, Environment and Biotechnology. 2020;**61**:51-59. DOI: 10.1007/s13580-019-00200-1

[73] Glenn GM, Poovaiah BW. Cuticular properties and postharvest calcium applications influence cracking of sweet cherries. Journal of the American Society for Horticultural Science. 1989;**114**(5):781-788

[74] Tony W, John C. All about cherry cracking. Tree Fruit Leader. 1994;**3**(2)

[75] Mignani I, Greve IC, BenArie R, Stotz HU, Shockel K, Labavitch J. The effects of GA_3 and divalent cations on aspects of pectin metabolism and tissue softening in ripening tomato pericarp. Physiologia Plantarum. 1995;**93**:108-115. DOI: 10.1034/j.1399-3054.1995.930116.x

[76] Jackman RL, Stanley DW. Perspectives in the textural evaluation of plant foods. Trends in Food Science & Technology. 1995;**6**(6):187-194

[77] Jifon JL, Lester GE. Foliar potassium fertilization improves fruit quality of field-grown muskmelon on calcareous soils in South Texas. Journal of Science and Food Agriculture. 2009;**89**:2452-2460

[78] Lopez-Zaplana A, Bárzana G, Ding L, Chaumont F, Carvajal M. Aquaporins involvement in the regulation of melon (*Cucumis melo* L.) fruit cracking under different nutrient (Ca, B and Zn) treatments. Environmental and Experimental Botany. 2022;**201**:104981

[79] Davarpanah S, Tehranifar A, Abadia J, Val J, Davarynejad G, Aran M, et al. Foliar calcium fertilization reduces fruit cracking in pomegranate (*Punica granatum* cv. Ardestani). Scientia Horticulturae. 2018;**230**:86-91. DOI: 10.1016/j.scienta.2017.11.023

[80] Jifon JL, Syvertsen JP. Kaolin particle film applications can increase photosynthesis and water use efficiency of ruby red grapefruit leaves. Journal of the American Society for Horticultural Science. 2003;**128**:107-112

Chapter 6

Pumpkin Seeds Germination and Seedling Growth under Abiotic Stress

Masoumeh Asadi Aghbolaghi, Mohammad Sedghi, Ghasem Parmoon and Beata Dedicova

Abstract

An increase in heavy metals (HMs) pollution due to agricultural and industrial activities has become a serious environmental problem. Heavy metal poisoning and its accumulation in food chains are one of the major environmental and health problems of modern societies. Heavy metal poisoning and its accumulation in food chains are major environmental and health risks. Among these metals, lead and cadmium are the metals with the most concern due to their toxicity potential for animals and plants even for humans. The effect of seed aging on the germination and growth of *Cucurbita pepo* L. seedlings has been evaluated in an experiment with a completely random design where seeds were exposed to deterioration at 45°C and 95% humidity for 2, 4, 6, 8, 10, 12, 14, 16, 18, and 20 days. The measured characteristics included germination percentage and speed, germination uniformity, and days to reach 10 and 50% of germination. Changes in seed viability were evaluated by using the tetrazolium test. The results of the analysis of variance showed that aging had a significant effect on all studied characteristics at the level of 1%. It was also observed that during aging, the amount of color accepted by the seed tissue is decreasing, which indicated a decrease in the vigor and viability of seeds.

Keywords: *C. pepo*, accelerating aging, tetrazolium test, heavy metals, cadmium

1. Introduction

An increase in heavy metals (HMs) pollution due to agricultural and industrial activities has become a serious environmental problem in the current world [1–3]. Heavy metal poisoning and its accumulation in food chains are major environmental and health problems in modern societies [3–5]. Among these metals, lead and cadmium are the metals with the most concern due to their toxicity potential for animals and plants [6, 7]. The most significant effects of heavy metal toxicity are oxidative stress, pigment dysfunction, and change in protein activity [8].

Plants are equipped with antioxidant defense systems to eliminate or reduce oxidative damage [9]. The plant antioxidant system is composed of antioxidant enzymes such as superoxide dismutase (SOD), ascorbate peroxidase (APX), guaiacol peroxidase (GPX), and catalase (CAT) [10]. It has been shown that high cadmium concentrations disrupt most physiological processes in plants [11]. As cadmium concentration increased, there is a significant increase in catalase and peroxidase. As it was shown in our previous study There have been reports that pumpkin tends to collect small amounts of heavy metals [12] and also showed that concomitant use of humic acid reduced the destructive effects of heavy metals on the *C. pepo* L. by reducing cadmium uptake by seeds.

C. pepo L. is one of the most important crops that belong to the *Cucurbitaceae* family, an herbaceous, perennial, and polymorphic vegetable that grows in tropical conditions [13, 14]. Pumpkin fruit contains large amounts of vitamins, minerals, and biologically active substances [15]. The pumpkin seed oil contains high amounts of unsaturated fatty acids and is effective in treating intestinal worms, prostate hypertrophy, stomach, intestinal inflammation, and atherosclerosis lowers LDL levels, prevents irregular heart contractions, reduces common blood clots, and reduces the risk of bladder and kidney stone formation [16, 17].

Medicinal plants are the main reservoirs of many medicinal compounds and substances, which are influenced by environmental factors in addition to genetic factors. *C. pepo* L. is one of the valuable medicinal plants in the pharmaceutical industry and its oil is used in medicine. The most important components of the oil of this plant are linoleic and oleic fatty acids, sterols, micronutrients, vitamins, and carotenoids [18].

Seedling germination and growth are one of the most important stages of plant growth, which determines the degree of success of agricultural systems [19]. These stages are strongly influenced by seed quality (viability and seed vigor) [20]. Temperature, relative humidity of the environment, and seed humidity are the main factors in the preservation and viability of seeds during storage [21]. Increasing the amount of seed moisture causes an increase in the aging speed [22] therefore if the temperature and relative humidity of the environment are high, the seeds will deteriorate sooner and they will be closer to death while reducing the quality. The reduction of plasma membrane integrity, molecular changes in the structure of nucleic acids and stimulation of lipid peroxidation, and reduction of the activity of hydrolytic enzymes are among the most important changes that occur in the seed during deterioration. These changes can lead to a decrease in seed quality, percentage, and speed of germination, and slower growth of seedlings [23].

In the deteriorated seeds due to disturbances in cellular organelles such as mitochondria and glyoxysomes, the production rate of reactive oxygen species and superoxide radicals increases, which causes damage to cells [24]. The tetrazolium test is used to estimate the viability of seeds. In this test, dead and alive tissues are distinguished based on their relative respiration rate in the condition that they have absorbed water.

In the tetrazolium test, seed viability is determined based on the activity level of dehydrogenase enzymes. The dehydrogenase enzyme reacts with the precursors during respiration and causes the release of hydrogen ions into the oxidized and colorless salt solution. These ions are combined with tetrazolium and turn this colorless salt into red formazan salt. The continuation of the viability of the seeds is determined based on the staining pattern of the embryo and the intensity of its staining [25]. Therefore, we use the tetrazolium test to study the viability and aging process of pumpkin seeds as a complementary test to our previous study with accumulations of cadmium in pumpkin seeds and the effect of humic acid.

2. Materials and methods

2.1 Seeds germination and tetrazolium test

Pumpkin seeds for this experiment were obtained from Pakan Bazr Isfahan Company, Isfahan, Iran (32°38′41”N51°40′03″ E).

This experiment was carried out in the form of a completely random design with three repetitions. In order to perform aging, an accelerated aging test was used. In this method, the seeds were kept in a mesh container in a seeds germinator (RICO, Scientific Instruments, Germany) with a relative humidity of 95% and with a temperature of 45°C for 2, 4, 6, 8, 10, 12, 14, 16, 18, and 20 days. To carry out the germination test, the seeds were disinfected with 1% sodium hypochlorite for 3 minutes and then 25 seeds were placed in a 9 cm Petri dish with two layers of Whatman paper and were transferred to the germinator with a temperature of 25°C. Counting of germinating was done daily for 14 days and the criterion of germinating was 2 mm of the root.

The Germinator software was used to calculate the trait characteristics related to seed germination. This software calculates Gmax (maximum germination), T50 (time required to reach 50% germination) and T10 (time required to reach 10% germination), CU (uniformity of germination = the time interval for germination rate to reach from 25–75%) and AUC (germination speed). Regarding germination uniformity, the lower the number obtained, the more uniform the germination of the seeds is [26].

The tetrazolium test was used to evaluate the seed viability and vigor. For this purpose, seeds were first soaked in distilled water for 24 hours to increase their respiratory activity. Then, the seeds were kept for 2 hours in 5 ml of 1% tetrazolium solution (2,3,5triphenyl tetrazolium chloride (TTC)) with 0.05 M phosphate buffer (pH 3.7) and then their viability and dyeability were evaluated [27].

Analysis of data was done using SAS 9.2 software and a comparison of means was done using an LSD test at a 5% probability level. The graphs were drawn with Excel.

3. Results and discussion

Germination ($P \leq 0.1$), (**Table 1**). Aging caused a decrease in the percentage of germination. So that in the conditions of nonaging and up to 2 days after aging, pumpkin seeds germination was 92%. However, with the increase in seed, the results

		MS				
SOV	df	Gmax	T50	T10	CU	AUC
Accelerated aging	10	0/17**	2568/57**	4327/9**	939/28**	521/39**
Error	22	0/0005	13/96	35/47	19/73	1/95
CV (%)	—	13/7	13/2	7/37	9/72	9/66

***Significant at the 1% level p < 0.01 probability level. Gmax: max of germination, T10 and T50 days: 10 and 50% of germination; CU: germination uniformity, AUC: speed of germination. SOV: the source of variation; MS: the mean square; CV: the coefficient of variation.*

Table 1.
Analysis of the variance of pumpkin seeds aging.

of the analysis of variances showed that accelerated aging had an effect on the maximum aging time, germination decreased and reached its minimum value of 29% in 20 days of aging (**Figure 1**). Gholami and Benny [15] also reported a reduction of germination during seed deterioration. During the deterioration, the production of reactive oxygen species, causing lipid peroxidation and damage to the cell membrane, and the increase of released fatty created a disturbance in the process of seed imbibition, which causes an increase in the leakage of seed reserves and a decrease in germination [28]. Govender et al. [29] also showed that corn seed storage for 1 year under natural conditions caused a decrease in germination percentage. They stated that the reason could be the presence of disease-causing fungi in natural storage conditions. It has been shown in research that the activity of enzymes, especially hydrolytic enzymes, decreases during seed aging, which can cause a decrease in germination [30], and also reported a reduction of germination during seed deterioration. During the deterioration, the production of reactive oxygen species, causing lipid peroxidation and damage to the cell membrane, and the increase of released fatty created a disturbance in the process of seed imbibition, which causes an increase in the leakage of seed reserves and a decrease in germination [31]. Govender et al. [32] showed that corn seed storage for 1 year under natural conditions caused a decrease in germination percentage. They stated that the reason could be the presence of disease-causing fungi in natural storage conditions. It has been shown in research that the activity of enzymes, especially hydrolytic enzymes, decreases during seed aging, which can cause a decrease in germination [19]. It has also been reported that deterioration causes a decrease in seed vigor and can disrupt the activity of antioxidant enzymes in pumpkin seeds [15].

The germination period was also influenced by seed aging, so aging at the level of 1 percent was significant in the time required to reach 10 and 50 percent germination (**Table 1**). The comparison of the means showed that accelerated aging for 20 days caused an increase in the time of germination (**Figures 2** and **3**). The highest rate of germination about time was obtained in the control treatment, and its value reached its lowest value from thy 16 of deterioration (**Figure 4**). Being high in this quality, in addition to showing high germination, is also an expression of high germination

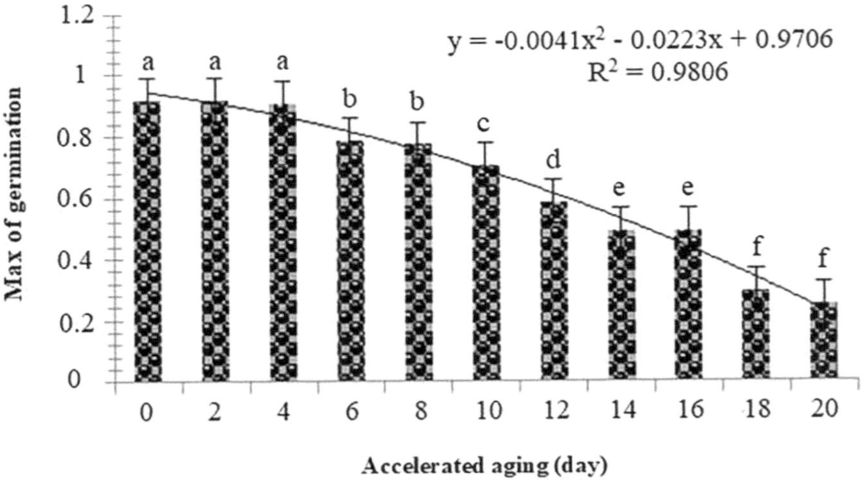

Figure 1.
Effect of accelerated aging on the maximum germination of pumpkin seed.

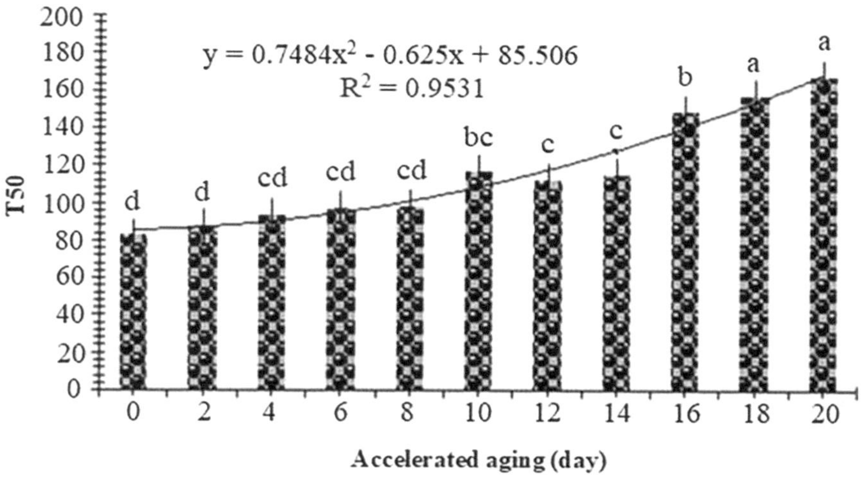

Figure 2.
Effect of accelerated aging on T50 (day to 50% germination) of pumpkin seed.

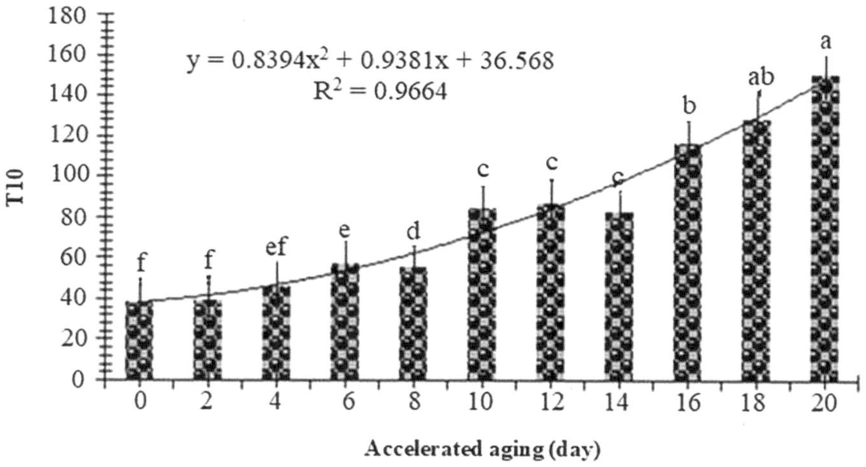

Figure 3.
Effect of accelerated aging on T10 (day to 10% germination) of pumpkin seed.

speed. Accelerated aging at the level of 1 percent also had a significant difference in the uniformity of germination (**Table 1**). In general, it can be said that accelerated aging caused a decrease in the uniformity of germination (**Figure 5**).

A high germination time indicates a lower germination speed, which indicates low seed quality. The faster emergence of crops on the farm, the better the establishment of the seedlings. High germination speed gives agricultural plants the possibility of better competition against weeds, reduces their damage, and increases the yield of agricultural products. Soltani et al. [33] showed that increasing the storage time of wheat seeds at a temperature of 40°C caused an increase in the average germination time. It has been said that the reason for the decrease in seed germination speed under high temperatures and temperatures is the loss of seed viability due to the loss of membrane health [34]. Veselova and Veslovesky [35] noted increasing the storage

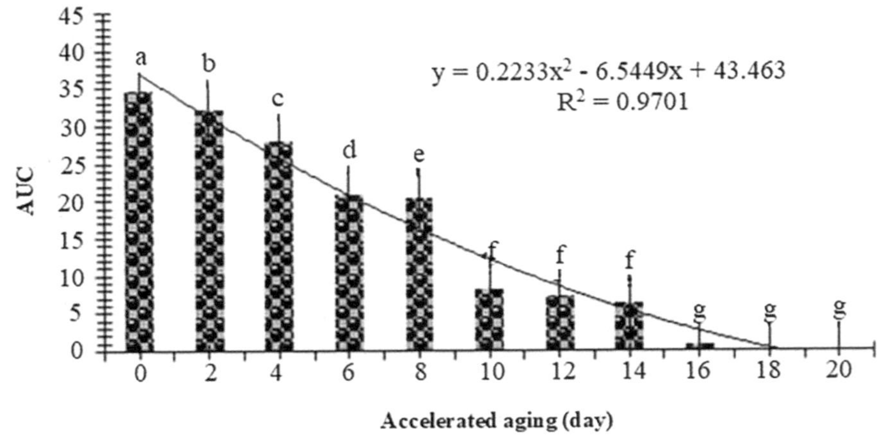

Figure 4.
The effect of accelerated aging on the germination speed of pumpkin seed.

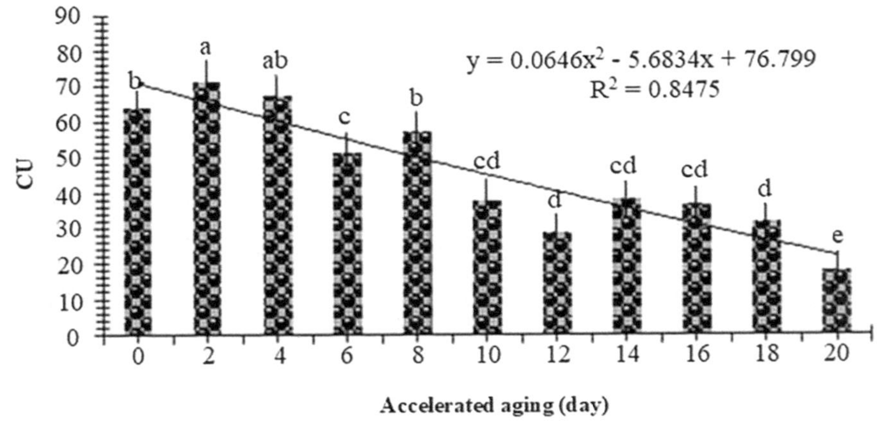

Figure 5.
The effect of accelerated aging on the germination uniformity of pumpkin seed.

time of seeds under the conditions of accelerated aging causes an increase in seed deterioration and a decrease in the percentage and speed of germination.

According to the results of biochemical tests, as a result of aging, the dyeability of the seed tissues is reduced, which indicates the reduction of seed viability and vigor. The results show that 2 days of aging did not affect seedling axis tissues in the embryo, but the differences with the increased aging time. At the lower level of aging, the first part whose dye ability decreases is part of the root and shoot. The increase in the aging time causes the expansion of the area of noncoloring to the center of the seed and cotyledons. The results showed that the dyeability of the seed tissues is the same as the results of the germination test, no significant frequencies were observed between the low levels of aging in terms of dyeability and germination.

At the beginning of germination, the rate of seed respiration increases by absorbing water during the imbibition process. An increase in respiration causes an increase

Figure 6.
Effect of accelerated aging on the viability of pumpkin seed using Tetrazolium test. The size bar is 1 mm. A No aging, B is 2 days C is 4 days. D is 6 days, E 8 days, F 10 days, G is 12 days, H is 14 days, I 16 days, F 18 days, and K is 20 days of aging.

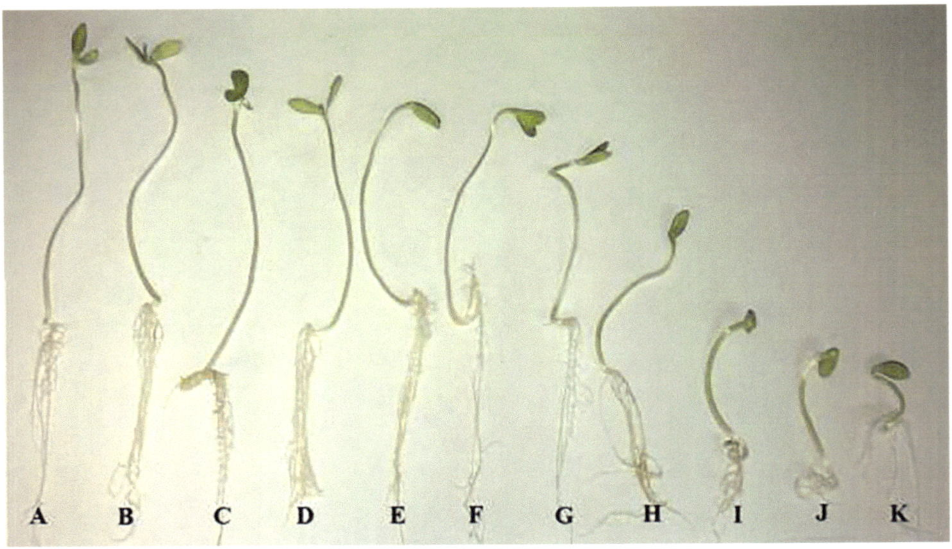

Figure 7.
Effect of accelerated aging on the growth of pumpkin seedlings (a: Non-aging, B: 2 days aging, C: 4 days aging, D: 6 days aging, E: 8 days aging, F: 10 days aging, G: 12 days aging, H: 14 days aging, I: 16 days aging, J: 18 days aging, K: 20 days aging).

in the reaction of the hydrogen produced during respiration with the hydrogenase enzyme, which causes the tetrazolium salt to create a colored compound that causes different parts of the seed to be colored. The higher the level of coloring, the higher the respiration in the seed and the higher the vigor and viability. Other studies have shown that during aging, the activity of hydrolytic enzymes decreases, which can cause a decrease in respiration [31]. As a result, the increase of H_2O_2 and free radicals in the cytoplasm of aging cells causes photosynthetic activities to become inactive. Protein binding decreases, and the sensitivity of proteins to proteolytic enzymes increases, which ultimately leads to a decrease in the viability of seeds [24].

Aging, in addition to the decrease in germination ability, caused a decrease in seed vigor. In fact, seed vigor is the first factor that decreases during aging [36]. The staining of the tissues during the tetrazolium test can be an indicator of seed vigor. The higher the coloring of the seed tissues, the higher the enzyme activity and the higher seed's vigor.

According to the results in **Figure 6** it can be seen that during the aging, the coloring of the seed tissues was reduced, which can be an indicator of the reduction of the vigor of this seed. The results of seedlings' growth also showed that their growth rate of them decreased due to aging. The decrease in the growth rate of seedlings can be caused by the decrease in seed vigor during aging. The growth of the seedlings showed that there is not much difference between the low levels of aging as well as the germination ability and seed survival and the length of their roots and stems was also the same, but the increase in the aging time caused a decrease in the growth rate of the seedlings (**Figure 7**). The length of the stems is one of the attributes that indicate a seed vigor. The seeds with low vigor may germinate, but due to the reduction in the length of the stem, they cannot emerge, and in this way, the percentage of emergence in the field is reduced. On the other hand, short stems have less emerging power due to their lower dry weight compared to long stems [23]. Soltani et al. [33] reported that the dry weight of seedlings decreased with increasing storage periods. The decrease in the dry weight of seedlings can be due to the decrease in the seed reserve or the decrease in the conversion efficiency of the dynamic reserve due to the decrease in the activity of hydrolytic enzymes during aging.

4. Conclusions

In conclusion, it can be confirmed that *C. pepo* L. in the tetrazolium test showed that aging causes a decreased percentage of germination, the uniformity of germination, and an increase in the average time required for germination, which ultimately causes a decrease in the viability of seeds. The decrease in the uniformity of germination caused the decrease in seed vigor due to aging, which ultimately causes a decrease in the growth of seedlings. Seed aging with the production of active oxygen species causes damage to the cell membrane and the activity of seed enzymes especially hydrolytic enzymes, which reduces the vigor of seeds and their viability.

Conflict of interest

The authors declare that they have no conflict of interest.

Author details

Masoumeh Asadi Aghbolaghi[1], Mohammad Sedghi[2], Ghasem Parmoon[3] and Beata Dedicova[4*]

1 Department of Agronomy and Plant Breeding, College of Agriculture and Natural Resources, University of Tehran, Karaj, Iran

2 Department of Agronomy and Plant Breeding, Faculty of Agriculture and Natural Resources, University of Mohaghegh Ardabili, Ardabil, Iran

3 Sugar Beet Research Department, Kermanshah, Agricultural and Natural Resources Research and Education Center, AREEO, Kermanshah, Iran

4 Department of Plant Breeding, Swedish University of Agricultural Sciences (SLU), Sweden

*Address all correspondence to: beata.dedicova@slu.se

References

[1] Imran MA, Sajid ZA, Chaudhry MN. Arsenic (As) toxicity to germination and vegetative growth of sunflower (*Helianthus annuus* L.). Polish Journal of Environmental Studies. 2015;**24**:1993-2002

[2] Khan MU, Shahbaz N, Waheed S, Mahmood A, Shinwari ZK, Malik RN. Comparative health risk surveillance of heavy metals via dietary foodstuff consumption in different land-use types of Pakistan. Human and Ecological Risk Assessment: An International Journal. 2016;**22**:168-186

[3] Rizwan M, Ali S, Qayyum MF, Ibrahim M, Zia-ur-Rehman M, Abbas T, et al. Mechanisms of biochar-mediated alleviation of toxicity of trace elements in plants: A critical review. Environmental Science and Pollution Research International. 2016;**23**:2230-2248

[4] Khan MJ, Jones DL. Effect of composts, lime and diammonium phosphate on the phytoavailability of heavy metals in a copper mine tailing soil. Pedosphere. 2009;**19**:631-641

[5] Gupta SK, Scott C, Mitra A. Advances in Land Resource Management for 21st Century. New Delhi, India: Soil Conservation Society of India; 2011. pp. 446-46

[6] Salazar MJ, Rodriguez JH, Nieto GL, Pignata ML. Effects of heavy metal concentrations (Cd, Zn, and Pb) in agricultural soils near different emission sources on quality, accumulation and food safety in soybean [*Glycine max* (L.) Merrill]. Journal of Hazardous Materials. 2012;**233**:244-253

[7] Zhao Y, Fang X, Mu Y, Cheng Y, Ma Q, Nian H, et al. Metal pollution (Cd, Pb, Zn, and As) in agricultural soils and soybean, glycine max, in southern China. Bulletin of Environmental Contamination and Toxicology. 2014;**92**:427-432

[8] Rastagoo L, Alemzadeh A, Tale AM, Tazangi SE, Eslamzadeh T. Effects of copper, nickel, and zinc on biochemical parameters and metal accumulation in gouan,' Aeluropus littoralis'. Plant Knowl. J. 2014;**3**:42-49

[9] Petrov VD, Van Breusegem F. Hydrogen peroxide–a central hub for information flow in plant cells. AoB Plants. 2012;**14**:1-13

[10] Michalak A. Phenolic compounds and their antioxidant activity in plants growing under heavy metal stress. Polish. Journal of Environmental Studies. 2006;**15**:523-530

[11] Tito GA, Chaves LHG, Fernandes JD, Monteiro DR, de Vasconcelos ACF. Effect of copper, zinc, cadmium, and chromium in the growth of crambe. Agricultural Sciences. 2014;**5**:975-983

[12] AghbolaghiM A, Sedghi M, Seyed Sharifi R. Dedicova B: Germination and the biochemical response of pumpkin seeds to different concentrations of humic acid under cadmium stress. Agriculture. 2022;**12**:1-15

[13] Sharma OP. Plant Taxonomy. New York City, United States: Tata McGraw-Hill Education. Pvt. Ltd.; 2010. p 553

[14] Tartoura EAA, El-Gamily EI, El-Waraky YBA, Kamel MKY. Effect of phosphorus fertilization and fruit thinning on seed production of summer squash plants 1-vegetative traits and leaves chlorophyll constituent. Journal of Plant Production. 2014;**5**:1807-1816

[15] Asadi, Aghbolaghi M, Razavi F. Improving germination and biochemical aspects in pumpkin (*Cucurbita pepo*) seeds deterioration under priming by salicylic acid and ascorbic acid. Iranian Journal of Seed Science and Technology. 2021;**10**:75-88

[16] Fruhwirth GO, Hermetter A. Production technology and characteristics of Styrian pumpkin seed oil. European Journal of Lipid Science and Technology. 2008;**110**:637-644

[17] Danilcenko H, Jariene E, Gajewski M, Cerniauskiene J, Kulaitiene J, Sawicka B, et al. Accumulation of elements in some organicallygrownalternativehorticultural crops in Lithuania. Acta Scientiarum Polonorum Hortorum Cultus. 2011;**10**:23-31

[18] Yadav M, Jain S, Tomar R, Prasad GBKS, Yadav H. Medicinal and biological potential of pumpkin: An updated review. Nutrition Research Reviews. 2010;**23**:184-190

[19] Soltani A, Robertson MJ, Torabi B, Yousefi-Daz M, Sarparast R. Modeling seedling emergence in chickpeas as influenced by temperature and sowing depth. Agricultural and Forest Meteorology. 2006;**138**:156-167

[20] De Figueiredo E, Albuquerque MC, De Carvalho NM. Effect of the type of environmental stress on the emergence of sunflower (*Helianthus annum* L.), soybean (*Glycine max* L.) and maize (*Zea mays* L.) seeds with different levels of vigor. Seed Science and Technology. 2003;**31**:465-479

[21] McDonough CM, Floyd CD, Waniska RD, Rooney LW. Effect of accelerated aging on maize, sorghum, and sorghum meal. Journal of Cereal Science. 2004;**39**:351-361

[22] Eisvand HR, Alizadeh MA, Fekri A. How hormonal priming of aged and non-aged seeds of brome grass affects seedling physiological characters. Journal of New Seeds. 2010;**11**:52-64

[23] Khajeh-Hosseini M, Powell AA, Bingham IJ. The interaction between salinity stress and seed vigor during germination of soybean seeds. Seed Science Technology. 2003;**31**:715-725

[24] Kibinza S, Vinel D, Come D, Bailly C, Corbineau F. Sunflower seed deterioration as related to moisture content during aging, energy metabolism, and active oxygen species scavenging. Plant Physiology. 2006;**128**:496-506

[25] Franca-Neto J, Kryzanonwski FC. Tetrazolium: An important test for physiological seed quality evaluation. Journal of Seeds Science. 2019;**41**:359-366

[26] Soltani A, Zeinali E, Galeshi S, Latifi N. Genetic variation for and interrelationships among seed vigor traits in wheat from the Caspian Sea coast of Iran. Seed Science and Technology. 2001;**29**:653-662 (In Persian)

[27] Lester R, Smith AL. The mode of reduction of tetrazolium salt by beef heart mitochondria; the role of coenzyme Q and other lipids. Biochemical Biophysical Act. 1961;**47**:475-496

[28] Fath A, Bethke P, Beligni V, Jones R. Active oxygen and cell death in cereal aleurone cells. Journal of Experimental Botany. 2002;**53**:1273-1282

[29] Sneideris LC, Gavassi MA, Campos ML, D'Amico-Damiao V, Carvalho RF. Effects of hormonal priming on seed germination of pigeon pea under cadmium stress. Anais da Academia Brasileira de Ciências. 2015;**87**:1847-1852

[30] Gholami Tile Benny H, Salehi Balashhry M, Farhady R. Priming effect, and deterioration of seed germination and seedling growth changes of rice (*Oryza sativa* L). Seed Science and Technology. 2013;**1**:1-13 (In Persian)

[31] Basra SMA, Ahmad N, Khan MM, Iqbal N, Cheema MA. Assessment of cotton seeds deterioration during accelerated aging. Seed Science and Technology. 2003;**31**:531-540

[32] Govender V, Aveling TAS, Kritzinger Q. The effect of traditional storage methods on germination and vigor of maize (*Zea mays* L.) from northern Kwa Zulu-Natal and southern Mozambique. South African Journal of Botany. 2007;**74**:190-196

[33] Soltani A, Kamkar B, Galeshi S, Akram Ghaderi F. Effect of seed storage on resource depletion and heterotrophic growth of the wheat seedling. Iranian journal of agriculture. Science. 2008;**15**:229-259

[34] Krishnan P, Nagarajan S, Moharir AV: Thermodynamic characterization of seed deterioration during storage under accelerated aging conditions. Biosystems Engineering. 2004;**89**:425-433

[35] Veselova TV, Veslovsky VA. Investigation of typical germination changes during accelerated aging of pea seeds. Seed Science and Technology. 2003;**31**:517-530

[36] McDonald MB. Seed deterioration. Physiology, repair, and assessment. Seed Science and Technology. 1999;**27**:177-237

Section 4

Molecular Tools for Studies on *Cucurbitaceae* Crops

Chapter 7

Plant Virus-Based Tools for Studying the Function of Gene and Noncoding RNA in Cucurbits

Ling-Xi Zhou, Xiang-Dong Li and Chao Geng

Abstract

Cucurbits are economically important crops worldwide. The genomic data of many cucurbits are now available. However, functional analyses of cucurbit genes and noncoding RNAs have been impeded because genetic transformation is difficult in many cucurbitaceous plants. The cucurbits-infecting plant viruses can be modified into useful tools for functional genomic studies in cucurbits, which provide alternative ways for rapid characterization of gene and noncoding RNA functions. This review will focus on the advancement and application of plant viruses-based gene silencing, gene expressing, and noncoding RNA regulation tools for studying the development, fruits, and stress of cucurbits. The features, advantages, and disadvantages of different plant virus vectors will be discussed in detail. We hope this review will provide guidance for studies on cucurbitaceous plants.

Keywords: plant virus vector, VIGS, VbMS, TRSV, *Potyvirus*, *Tobamovirus*, ALSV, cucurbits

1. Introduction

Cucurbitaceae is one of the most genetically diverse plant families worldwide, which is viewed as the second-largest fruit and vegetable family after *Solanaceae* [1]. Biotechnological advances could empower the molecular characterization of functional genes in cucurbits and will inform new methods of breeding cucurbit species [2]. In recent years, intensive high-throughput genomic sequencing studies have made considerable advances in molecular phylogeny and genomics. The completion of genome sequencing of several cucurbitaceous crops increases the need for rapid gene function analysis tools [1, 3, 4].

Traditional transgenic approaches and CRISPR/Cas9-mediated genome editing technologies accelerated gene function characterization and crop improvement [5], while most economically important cucurbit crops are difficult to transform, or the transformation efficiency remains very low [2]. Since plant virus vectors could systemically infect host plants, fragments or the full length of the target genes of interest would be delivered into the whole plants to modulate gene expression within a short time by using simple operations [6]. Now, plant virus-mediated gene delivery systems have been successfully employed as alternative biotechnology tools for gene function studies, particularly in plant species recalcitrant for genetic transformation, including cucurbit plants [7]. Viruses can replicate in plant cells, and offer numerous advantages in gene overexpression, including their maximum levels of multiplication and concomitant levels of transient gene expression from viral genomes [8]. In the past years, several plant virus-based protein expression vectors have been widely used in gene overexpression to understand their functions. Furthermore, the discovery of posttranscriptional RNA silencing (PTGS) [9] and the development of modern sequencing tools spurred the development of virus-induced gene-silencing (VIGS) screens to knock down the specific host genes, which accelerated advances in investigating their functions [10, 11]. In addition, VIGS can also be used to study the roles of particular genes in metabolic pathways [12]. In plants, *phytoene desaturase* (*PDS*) is involved in the carotenoid biosynthesis pathway, and Knockdown of this gene results in albino phenotypes due to the absence of chlorophylls [13], making it widely used as an indicator for VIGS in plants.

MicroRNAs (miRNAs) are essential noncoding riboregulators of gene expression in plants, which influence the development and physiology of plants and responses to biotic and abiotic stresses [14]. Although much progress has been made in revealing miRNA functions in some model plans, their roles are still incompletely understood, especially in crop plants [15]. Compared with constitutive expression in transgenic plants, viral vectors have advantages in production and potentially rapidly. In the past years, plant viruses have been engineered into virus-based miRNA silencing (VbMS) vectors to inhibit the miRNA function in many plants, which are suitable for high throughput of analyzing miRNA function [16–18].

Recently, several plant viruses have been harnessed for CRISPR/Cas9-based genome editing of plants, which could both express Cas9 protein and deliver single-guide RNA (sgRNA) for genome editing [19–22]. This type of virus-based strategy for gene editing is termed virus-induced genome editing (VIGE). Compared with traditional transgenic approaches, the VIGE process is more efficient and faster because the virus replication would give rise to high yields of sgRNAs and Cas proteins [23, 24].

Above all, the development of plant virus-based vectors has great implications for plant functional genomic studies. Different strategies are required for constructing virus vector-based systems due to their expression strategies and biological limitations [12], which lead to different suitable scenarios for the application of virus vectors. This review will discuss several plant virus-based tools used in cucurbits, including their construction strategies and examples of their applications.

2. Tobacco ringspot virus

Tobacco ringspot virus (TRSV) is the most well-characterized species of the genus *Nepovirus* in the family *Comoviridae*, which consists of a bipartite positive-strand RNA genome. Each RNA encodes a large polyprotein, which is proteolytically processed into mature proteins that encode different functions [25]. RNA1 is usually 8.1 to 8.4 kilobase (kb) nucleotides (nt) in length and encodes replication-related proteins, while RNA2 with a length from 3.4 to 7.2 kb nt usually encodes the capsid (CP) and movement protein (MP). TRSV can infect a wide range of plants, including cucurbits.

Zhao and co-authors developed TRSV virus-based vectors for foreign gene expression and VIGS in 2016. In the past few years, the application of viral 2A peptide allows the co-expression of multiple proteins from a single open reading frame (ORF), which prevents homologous recombination due to duplicated sequences for the protease cleavage sites, improving the stability of the vector [26]. In this research, they developed a vector with the 2A sequence upstream of the CP coding region to express green fluorescence protein (GFP). The 2A peptide could cleave the GFP protein from the polyprotein to produce partially free GFP. In addition, TRSV-based vectors showed the recovery phenotype, so GFP expression would decrease in the late infestation period [27]. To overcome this drawback, Fang et al. engineered TRSV vectors co-expressed GFP and heterologous viral suppressors of RNA silencing (VSRs) separated through two different 2A peptides in tandem, which produced stronger and more stable GFP expression in plants [16].

However, symptom recovery could be an advantage in the VIGS system for gene silencing experiments as viral symptoms usually confuse the silencing phenotypes. To develop the TRSV-based VIGS vector, researchers created a cloning site downstream of the CP stop codon for inserting the silencing gene sequence. This TRSV-based VIGS system caused clear silencing phenotypes and long duration of VIGS phenotype in cucurbits [16, 27]. Since TRSV-based VIGS vector was developed, it has been increasingly used in cucurbit crop studies such as the molecular mechanisms of defense-related genes and the multicellular trichome development due to its excellent silencing efficiency [28, 29].

In 2021, Fang et al. further developed the TRSV vector used for studying the function of miRNA in cucurbits. The phenotypes induced by TRSV-based miRNA silencing vector are obvious, and the efficiency of most TRSV-based miRNA silencing is high in both model plants and cucurbit plants. The silencing efficiency of miRNA was about 75.0–87.6% in loofah (*Luffa aegyptiaca*), and 68.8–75.0% in melon(*Cucumis melo*), which provides potential solutions for the functional study of cucurbit miRNAs [16]. It is noteworthy that this is the first report for the VbMS vector for the miRNA silencing in cucurbit plants.

3. Tobamoviruses

The genus *Tobamovirus* belongs to the family *Virgaviridae,* tobamoviruses assemble rod-shaped particles that are approximately 300–310 nm long and 18 nm wide and encapsidate a single-stranded, positive-sense RNA of 6.3–6.6 kb nt in length as its genome. Tobamoviruses can easily yield a large number of viral particles within a short time after they infect host plants [30], so these viruses can be used as good tools for the overexpression of genes [31]. Tobacco mosaic virus (TMV), the most well-characterized member of the genus *Tobamovirus,* has been extensively used for assessing gene function [32]. Tobamoviruses encode a 130-kDa protein (small replicase subunit), a 180-kDa readthrough protein (RNA-dependent RNA polymerase), an MP, and a CP. MP and CP were expressed *via* subgenomic RNAs (sgRNA) during infections [33]. Researchers showed that the most highly expressed viral gene is controlled by the CP subgenomic promoter (SGP) [34]. To establish a virus-based expression system, the foreign genes could replace CP. However, the recombinant viruses with genes only multiply in inoculated leaves [35]. It could be a good choice to place the heterologous gene at the terminal end of CP to express as a fusion coat protein if the protein expressed is small [36, 37]. The foreign genes could also be inserted in frame between *MP* and *CP* sequences, and the SGP in the *CP* gene was used to drive foreign gene expression. In the meanwhile, to prevent homologous recombination, CP SGP of other tobamoviruses was usually used to drive CP expression [38]. It is noteworthy that the length of the SGP should be empirically studied, as SGP length could affect virus accumulation and gene expression levels in plants [39]. TMV vectors designed with a duplication of the TMV 3′-UTR between the foreign ORF and the *CP* gene could increase the expression of the foreign gene [40].

3.1 Cucumber green mottle mosaic virus

Cucumber green mottle mosaic virus (CGMMV), which mainly infects cucurbits under natural conditions, is a member of the genus *Tobamovirus*, CGMMV has been extensively exploited as suitable gene silencing and overexpression vectors in cucurbits [39]. In the early period, CGMMV was used to express the small protein in muskmelon. The expression of dengue virus type 2 envelope (E) protein and hepatitis B surface antigen (HBsAg) was achieved by inserting into the multiple cloning sites to cover the stop codon of CP and resulted in the expression of the target gene as a readthrough CP-fused protein. However, the T7-promoter-driven infectious clone used in these studies needs *in vitro* transcription to obtain the viral RNA transcripts then mechanically inoculated into muskmelon, which makes it inconvenient to use [36, 37]. In 2015, Zheng et al. developed a CGMMV infectious clone tagged with GFP under the transcriptional control of the CaMV 35S promoter, which can establish infection in cucumber (*Cucumis sativus*), watermelon (*Citrullus lanatus*), bottle gourd (*Lagenaria siceraria*), and *Nicotiana benthamiana* plants *via* agroinoculation [41]. The GFP coding region was placed in frame between the coding sequences between MP and CP. They chose 234 nt before the CP initiation codon as the CP 5′ subgenomic promoter based on their rapid amplification of cDNA ends (RACE) result and the sequences of CP SGP of TMV and constructed three vectors with different lengths of the 5′-proximal CP ORF to drive GFP expression, which successfully infected bottle gourd and displays strong green fluorescence in symptom leaves. However, GFP was easily deleted from the viral genome due to the recombination of duplicated SGP along with infection [41].

Shortly after that, Liu et al. developed a CGMMV-based VIGS vector that silences *PDS* in watermelon, melon, cucumber, and bottle gourd. They adopted a similar strategy with TMV VIGS vectors previously reported to generate a series of CGMMV VIGS vectors, which explored different lengths of CP SGP. Their results indicated that the *PDS* gene fragments induced a more robust silencing phenotype when the vector contained a 190-bp duplicated copy of the putative CP SGP. The length and structure of the insert sequence could affect the silencing efficiency. Their results showed that the CGMMV vectors harboring the *PDS* gene sequence of 300 bp in length had the highest silencing efficiency in cucurbits, and the silencing phenotypes could persist for at least one month [42]. Up to now, the CGMMV-based VIGS system has been widely used in cucurbits studies, such as the tolerance to salinity stress, chilling tolerance, and functional analyses of resistance-related genes [43–45].

3.2 Cucumber fruit mottle mosaic virus

Cucumber fruit mottle mosaic virus (CFMMV), which is also a member of the genus *Tobamovirus*, mainly infects cucurbits. As CGMMV, CFMMV has also been constructed as virus vectors for functional genomic studies in cucurbits. In 2016, Rhee et al. constructed a CFMMV vector, which could express *enhanced green fluorescent protein* (EGFP) in cucumber, melon, and watermelon. They tested the active region of CP SGP and found that the region from −55 to +100 nt was identified as the active core promoter. Then, they chose CP SGP with 93 nt before the CP initiation to express the CP of CFMMV, and CP SGP with 100 nt to drive the EGFP expression. The EGFP can be expressed in hypocotyl and leaf veins, but the expression level is low in mesophyll cells. They also found that co-infiltration with the P19 RNA silence suppressor could enhance EGFP expression [38].

Recently, Rhee et al. further constructed an efficient CFMMV VIGS vector that exhibits high gene silencing efficiency and long-lasting *PDS-silenced* phenotype in cucurbit plants. The CFMMV VIGS vector adopts a similar strategy as the CFMMV overexpression vectors they previously developed. They found that the best vector to express EGFP was also the best VIGS vector to silence *PDS*. The vector could effectively silence the corresponding *PDS* gene in melon, cucumber, and watermelon, resulting in photobleaching in leaves and reproductive organs. It is worth noting that they used this optimized system to identify genes involved in male sterility, which suggests that VIGS in watermelon is functional [46].

Although some tobamoviruses vectors show significant overexpression and silencing effects in cucurbits soon after inoculation, the repeated CP SGP sequences would result in homologous recombination. In the future, the effect of CGMMV-based gene overexpression and silencing will be improved by replacing the native SGP of CGMMV CP with the SGP of other tobamoviruses CP to stabilize the inserted fragments.

4. Apple latent spherical virus

Apple latent spherical virus (ALSV) is a member of the genus *Cheravirus* (family *Comoviridae*), which is known for the presence of atypical members. ALSV contains two single-stranded RNA (ssRNA) species and three capsid proteins. RNA 1 has a single ORF encoding a polypeptide, which contains the consensus motifs of the protease cofactor, the NTP-binding helicase, the cysteine, protease, and the RNA polymerase. RNA 2 also has a single ORF encoding a polypeptide of 119 K/108 K containing a 53 K/42 K movement protein on the N-terminal side and three capsid proteins (Vp25, Vp20, and Vp24) [47].

In the early period, researchers constructed ALSV vectors to express GFP allowing us to trace the cell-to-cell and long-distance movement of ALSV in infected plant tissues of *N. benthamiana* [48, 49]. As ALSV is a kind of latent virus and does not induce any obvious symptoms in most host plants, it can be used for functional genomics in the host plants. This is one of the requirements for plant virus vectors used for VIGS [50]. ALSV was developed into VIGS vectors by Igarashi et al. in 2009. The target gene fragment can be inserted between 42KP and Vp25. The ALSV vector induced highly uniform phenotypes induced by knocking down *the subunit of magnesium chelatase* (*SU*) and *PDS* in plants, including cucurbit species such as cucumber, watermelon, zucchini (*Cucurbita pepo*), loofah, and bottle gourd. ALSV-based VIGS vector can induce a highly obvious and stable gene-silenced phenotype throughout plant growth in infected plants. Recently, they further developed an efficient virus-induced gene silencing system using ALSV in pumpkins (*Cucurbita spp.*) [51].

However, ALSV vectors have several limitations, such as the ALSV vector is difficult to use for high-throughput functional genomics since ALSV proteins are expressed by polyprotein proteolytic processing and the inserted gene sequences cannot produce stop codon in the open reading frame of ALSV. Another disadvantage is that ALSV-cDNA clones are less efficient for direct inoculation into host plants [52].

5. Potyviruses

Potyvirus is a genus that belongs to the family *Potyviridae*. Potyviruses exist as flexuous rod-shaped particles ranging from 720 to 900 nm in length with an ssRNA genome of approximately 10 kb. Potyviruses have only an ORF encoding a large polyprotein, which is processed into ten functional proteins by three viral proteinases [53]. P3N-PIPO protein is produced using a viral polymerase slippage mechanism [54]. Now, lots of potyviruses have been developed into infectious clones for potyviral studies expression vectors [54]. The foreign gene fragments are usually inserted between *the first protein* (*P1*) and *helper component-proteinase* (*HC-Pro*) or between *nuclear inclusion protein b* (*NIb*) and *CP* cistrons [55, 56]. Potyviruses are considered promising expression vectors since their proteolytic processing strategy of gene expression ensures that the foreign protein can be produced in equal amounts with viral proteins [57]. Compared with other known viral vectors that cause severe symptoms to host plants, attenuated mutants with a mutation in HC-Pro, which is both a major virulence determinant of potyviruses and RNA silence suppressor, make it possible for the potyviruses to be the environmentally safe virus vector for the expression of various foreign genes in plants [58].

5.1 Zucchini yellow mosaic virus

Zucchini yellow mosaic virus (ZYMV), which belongs to the genus *Potyvirus*, has been used as expression vector in cucurbit plants for a long time. In 2001, Arazi et al. exploited ZYMV to develop a novel virus-based vector system for the expression of foreign genes in cucurbits. The virus vectors in this study were generated from an attenuated mutant (AG) [57], which accumulates to the same virus accumulation level as the severe ZYMV strain in cucurbits, without eliciting any phenotypic and developmental impairment [59]. This vector could express GFP, uidA (β-glucuronidase; GUS), and human interferon-alpha 2 (IFN) in leaves, stems roots, male flowers, and fruits of squash (*Cucurbita spp.*) or cucumber. In the same year, they further expressed the *bar* gene using this vector in melon, cucumber, squash, and watermelon, which confer these plants' resistance against glufosinate ammonium-based herbicides [60]. In 2006, this vector was used both for overexpression- and down-regulation of *Trichoderma-induced MAPK* (*TIPK*) gene expression in cucumber [61]. In 2016, Kang and coauthors expressed GFP and bar in zucchini by simple rub-inoculation of plasmid DNAs of the ZYMV-based expression constructs [62].

5.2 Watermelon mosaic virus

Watermelon mosaic virus (WMV) is also a member of the genus *Potyvirus*, which exhibits a wide host range and is one of the most prevalent viruses in cucurbits worldwide [63]. In 2019, Aragonés and coauthors characterized a WMV isolate that induces mild symptoms in cucurbits [64], and then they developed a VIGS vector based on the mild isolate. Melon *PDS* fragments were inserted between *NIb* and *CP* in the form of sense, antisense, and hairpin. *PDS* fragments in sense and antisense orientations were more stable in the viral progeny, and the sense construct triggered the best silencing effect compared to those constructs with antisense and hairpin fragments. They further confirmed this result by lytargeting melon Magnesium chelatase subunit I (*CHLI*) using the VIGS vector. Silencing of *CHLI* can be easily tracked by a foliar decoloration phenotype, as a consequence of deficient chlorophyll accumulation, which exhibited decolored spots. The results showed that the VIGS vector constructed based on the mild isolate WMV could be a useful tool in cucurbit gene studies [65]. As with other potyviruses, WMV could also be used as an expression vector in cucurbits [58, 66].

Potyviruses are the largest group of plant RNA viruses, with approximately 200 species [67]. However, relatively few VIGS vectors have been derived from potyviruses, as potyviruses possess a strong RNA silencing suppressor that usually precludes VIGS. Some nonsynonymous mutants in HC-Pro can abolish HC-Pro's RNA silencing suppression (RSS) activity [67]. Therefore, a reliable potyvirus-based VIGS vector could be developed by introducing site-directed mutations to the *HC-Pro* gene, which attenuates virus virulence [65].

6. Tobacco rattle virus

Tobacco rattle virus (TRV) is one of the members of the genus *Tobravirus*. The genome of TRV is divided into two positive-sense, ssRNAs, each of which is encapsidated separately into rod-shaped particles. The genome of TRV RNA1 encodes replicase, 1a and 1b, which is sufficient for replication and movement within the host plants. The genome of RNA2 encodes CP, which can produce virus particles, and allows nematode-mediated transmission between hosts [68, 69]. TRV induces very mild symptoms and is suitable for gene silencing in most plants. The nonessential 29.4 k and 32.8 k genes of RNA2 can be replaced with an MCS for the insertion of the target gene sequence [68].

In 2019, a TRV-VIGS system was applied in cucumber using a new infection method with a special agroinfiltration solution, which provides an efficient and easy method for functional analysis of genes in cucumber [70]. *PDS* and *glycerol-3-phosphate 2-O-acyltransferase 6* (*GPAT6*) gene were successfully silenced with this VIGS system in cucumber. The silencing of *GATP6* resulted in the resistance to autotoxin stress mimicked by cinnamic acid (CA) in cucumber. This TRV-VIGS system enables efficient silencing of cucumber genes at the whole-plant level without time-consuming transformation or manipulation [71]. In the same year, Liao et al. developed transient overexpression and VIGS systems for oriental melon using the sprout absorption method. They successfully overexpressed and silenced candidate *lipoxygenases* (*LOXs*) gene in oriental melon using TRV-based VIGS they modified [72]. The overexpression of *CmLOX10* could trigger cell death responses in oriental melon. The disadvantage of the TRV-based VIGS vector is that, it's difficult to infect cucurbits *via* traditional inoculation method. The infection method used in these researches was established based on the whole cotyledonary node method of regeneration described by Ma and Wu [70, 73], which is operation complexity and time-consuming.

7. Cucurbit chlorotic yellows virus

Cucurbit chlorotic yellows virus (CCYV) is a recently discovered cucurbit-infecting virus, which belongs to the *Crinivirus* genus in the family *Closteroviridae*. CCYV has characteristic long, flexuous rod-shaped virions, ranging in size from 700 to 900 nm or 650–850 nm. The genome of CCYV contains two RNAs. RNA1 includes two ORFs, ORFs 1a and 1b, which encode papain-like cysteine proteinase (PRO), methyltransferase (MTR), helicase (HEL), and RdRp, respectively. RNA2 includes seven ORF that individually encodes CP (major coat protein), CPm (minor coat protein), 70-kDa heat shock proteins (Hsp70h), and P59, which are essential for virion structural components, movement, and vector transmission. CCYV causes chlorotic leaf spots and yellowing symptoms on cucurbit leaves [74]. Respectively, Wei et al. constructed a CCYV vector for GFP expression. They inserted different subgenomic RNA promoters and a GFP coding sequence into the upstream of the CP coding sequence to initiate the expression of GFP and CP expression. The GFP fluorescence was only detectable in cucumber leaf veins and surrounding cells [74]. As the infection of CCYV is limited in the phloem, CCYV-based vectors are specifically useful for studying the function of gene localized in the phloem [75].

8. Bean yellow dwarf virus

Bean yellow dwarf virus (BeYDV) belongs to the *Mastrevirus* genus of the family *Geminiviridae*. BeYDV comprises monopartite or bipartite circular single-stranded circular DNA (ssDNA) viruses characterized by their germinate particles comprised of two joined incomplete icosahedra [76]. The circular ssDNA of BeYDV is 2.5–2.7 kb in length and encodes MP, CP, and replication associated proteins A and B (RepA and RepB) [77]. BeYDV uses a rolling circle mechanism to replicate its genome, resulting in a very high yield of copies, which has been used to boost protein expression in transgenic plants and for efficient transient expression of foreign protein plants [78, 79]. BeYDV has a broad host range in plants, including many cucurbits. Yamamoto et al. developed a transient expression system that combines geminiviral replication and a double terminator, which could improve the protein expression in melon. In particular, the expression level was the highest when the HSP and Ext terminators were used as double terminators [80]. However, this vector cannot infect melon systemically, which could only be used for transient expression.

9. Conclusion

Most cucurbit plants possess very low transformation efficiency [81]. Plant virus-based tools have been used in gene silencing and overexpression in cucurbits. Up to now, TRSV and ALSV vectors, which have been used in different research groups, display good performance in cucurbit plants [16, 28, 29, 51, 82]. The efficiency of other virus vectors infecting cucurbit plants is relatively low. However, the VIGS capacity of CFMMV-based vectors has been enhanced recently, which provides guidance for the modification of other viral vectors [46]. In addition, VIGE technologies have been developed to overcome the bottleneck caused by tissue culture [24]. Some plant viruses mentioned in this review that infect cucurbits such as BeYDV and TRV have been harnessed for VIGE in *N. benthamiana*, *Arabidopsis thaliana*, potato (*Solanum tuberosum*), and tomato (*Solanum lycopersicum*) [83–86]. However, these VIGE tools have not been implemented in cucurbit crops yet. TMV has been constructed to deliver biologically functional single guide RNAs (sgRNAs) through the CP subgenomic promoter in *N. benthamiana* in 2017 [22]. As members of *Tobamovirus*, CGMMV and CFMMV also have the potential of delivering sgRNA. Because the correlation between the stability of virus vectors and the size of foreign genes is negative, it is difficult for most plant viruses to directly deliver large proteins such as Cas9 [87]. Up to now, most types of plant virus vectors were used to deliver gRNAs into Cas9-overexpressed plants for VIGE [24]. However, TRSV-based vectors, which have displayed powerful capacity to knock down genes and miRNAs in cucurbits, were recently reported to express SaCas9 in *N. benthamiana* without comprising virus stability [16]. Therefore, with the development of biology technology, plant virus-based tools will further facilitate the study of cucurbit plants.

Acknowledgements

This study was funded by National Natural Science Foundation of China (NSFC, 31801704), China Postdoctoral Science Foundation (2021 M691973; 2022 T150389), and Research and Development Program in Shandong Province (2021LZGC015).

Conflict of interest

The authors declare no conflict of interest.

Author details

Ling-Xi Zhou, Xiang-Dong Li and Chao Geng*
Department of Plant Pathology, Shandong Provincial Key Laboratory of Agricultural Microbiology, Col-lege of Plant Protection, Shandong Agricultural University, Tai'an, Shandong, China

*Address all correspondence to: chaogeng@sdau.edu.cn

References

[1] Guo J, Xu W, Hu Y, Huang J, Zhao Y, Zhang L, et al. Phylotranscriptomics in Cucurbitaceae reveal multiple whole-genome duplications and key morphological and molecular innovations. Molecular Plant. 2020;**13**(8):1117-1133

[2] Xin T, Tian H, Ma Y, Wang S, Yang L, Li X, et al. Targeted creation of new mutants with compact plant architecture using CRISPR/Cas9 genome editing by an optimized genetic transformation procedure in cucurbit plants. Horticulture Research. 2022;**9**:uhab086

[3] Bellot S, Mitchell TC, Schaefer H. Phylogenetic informativeness analyses to clarify past diversification processes in Cucurbitaceae. Scientific Reports. 2020;**10**(1):488

[4] Yu J, Wu S, Sun H, Wang X, Tang X, Guo S, et al. CuGenDBv2: An updated database for cucurbit genomics. Nucleic Acids Research. 2023;**51**(D1):D1457-D1D64

[5] Bhatta BP, Malla S. Improving horticultural crops via CRISPR/Cas9: Current successes and prospects. Plants (Basel). 2020;**9**(10):1360

[6] Choi B, Kwon S-J, Kim M-H, Choe S, Kwak H-R, Kim M-K, et al. A plant virus-based vector system for gene function studies in Pepper1. Plant Physiology. 2019;**181**(3):867-880

[7] Wang M, Gao S, Zeng W, Yang Y, Ma J, Wang Y. Plant virology delivers diverse toolsets for biotechnology. Viruses. 2020;**12**(11):1338

[8] Scholthof HB, Scholthof KB, Jackson AO. PLANT VIRUS GENE VECTORS FOR TRANSIENT EXPRESSION OF FOREIGN PROTEINS IN PLANTS. Annual Review of Phytopathology. 1996;**34**(1):299-323

[9] Lindbo JA, Silva-Rosales L, Proebsting WM, Dougherty WG. Induction of a highly specific antiviral state in transgenic plants: Implications for regulation of gene expression and virus resistance. The Plant Cell. 1993;**5**(12):1749-1759

[10] Kumagai MH, Donson J, Della-Cioppa G, Harvey D, Hanley K, Grill LK. Cytoplasmic inhibition of carotenoid biosynthesis with virus-derived RNA. Proceedings of the National Academy of Sciences of the United States of America. 1995;**92**(5):1679-1683

[11] Ruiz MT, Voinnet O, Baulcombe DC. Initiation and maintenance of virus-induced gene silencing. The Plant Cell. 1998;**10**(6):937-946

[12] Pogue GP, Lindbo JA, Stephen J, Garger A, Fitzmaurice WP. Making an ally from an enemy: Plant virology and the new agriculture. Annual Review of Phytopathology. 2002;**40**(1):45-74

[13] Vaia G, Pavese V, Moglia A, Cristofori V, Silvestri C. Knockout of phytoene desaturase gene using CRISPR/Cas9 in highbush blueberry. Frontiers in Plant Science. 2022;**13**:1074541

[14] Li M, Yu B. Recent advances in the regulation of plant miRNA biogenesis. RNA Biology. 2021;**18**(12):2087-2096

[15] Liao Q, Tu Y, Carr JP, Du Z. An improved cucumber mosaic virus-based vector for efficient decoying of plant microRNAs. Scientific Reports. 2015;**5**:13178

[16] Fang L, Wei XY, Liu LZ, Zhou LX, Tian YP, Geng C, et al. A tobacco ringspot virus-based vector system for gene and microRNA function studies in cucurbits. Plant Physiology. 2021;**186**(2):853-864

[17] Zhao J, Rios CG, Song J. Potato virus X-based microRNA silencing (VbMS) In potato. Journal of Visualized Experiments. 2020;**159**:1-9

[18] Sha A, Zhao J, Yin K, Tang Y, Wang Y, Wei X, et al. Virus-based microRNA silencing in plants. Plant Physiology. 2014;**164**(1):36-47

[19] Ariga H, Toki S, Ishibashi K. Potato virus X vector-mediated DNA-free genome editing in plants. Plant and Cell Physiology. 2020;**61**(11):1946-1953

[20] Yin K, Han T, Liu G, Chen T, Wang Y, Yu AYL, et al. A geminivirus-based guide RNA delivery system for CRISPR/Cas9 mediated plant genome editing. Scientific Reports. 2015;**5**(1):14926

[21] Ellison EE, Nagalakshmi U, Gamo ME, Huang P-j, Dinesh-Kumar S, Voytas DF. Multiplexed heritable gene editing using RNA viruses and mobile single guide RNAs. Nature Plants. 2020;**6**(6):620-624

[22] Cody WB, Scholthof HB, Mirkov TE. Multiplexed gene editing and protein overexpression using a tobacco mosaic virus viral vector. Plant Physiology. 2017;**175**(1):23-35

[23] Cody WB, Scholthof HB. Plant virus vectors 3.0: Transitioning into synthetic genomics. Annual Review of Phytopathology. 2019;**57**:211-230

[24] Oh Y, Kim H, Kim S-G. Virus-induced plant genome editing. Current Opinion in Plant Biology. 2021;**60**:101992

[25] Sanfaçon H. Nepovirus. In: Mahy BWJ, Van Regenmortel MHV, editors. Encyclopedia of Virology (Third Edition). Oxford: Academic Press; 2008. p. 405-413.

[26] Gopinath K, Wellink J, Porta C, Taylor KM, Lomonossoff GP, van Kammen A. Engineering cowpea mosaic virus RNA-2 into a vector to express heterologous proteins in plants. Virology. 2000;**267**(2):159-173

[27] Zhao F, Lim S, Igori D, Yoo RH, Kwon SY, Moon JS. Development of tobacco ringspot virus-based vectors for foreign gene expression and virus-induced gene silencing in a variety of plants. Virology. 2016;**492**:166-178

[28] Bartholomew ES, Xu S, Zhang Y, Yin S, Feng Z, Chen S, et al. A chitinase CsChi23 promoter polymorphism underlies cucumber resistance against Fusarium oxysporum f. sp. cucumerinum. New Phytologist. 2022;**236**(4):1471-1486

[29] Dong M, Xue S, Bartholomew ES, Zhai X, Sun L, Xu S, et al. Transcriptomic and functional analysis provides molecular insights into multicellular trichome development. Plant Physiology. 2022;**189**(1):301-314

[30] Hong Hanh Tran BC, Chen H, Menassa R, Hao X, Bernards M, Hüner NPA, et al. Developing a plant virus-based expression system for the expression of vaccines against porcine reproductive and respiratory syndrome virus. Journal of Virological Methods. 2019;**269**:18-25

[31] Scholthof KB. Tobacco mosaic virus: A model system for plant biology. Annual Review of Phytopathology. 2004;**42**:13-34

[32] Hefferon K. Plant virus expression vector development: New perspectives.

BioMed Research International. 2014;**2014**:785382

[33] Jiwan SD, White KA. Subgenomic mRNA transcription in Tombusviridae. RNA Biology. 2011;**8**(2):287-294

[34] Lehto K, Grantham GL, Dawson WO. Insertion of sequences containing the coat protein subgenomic RNA promoter and leader in front of the tobacco mosaic virus 30K ORF delays its expression and causes defective cell-to-cell movement. Virology. 1990;**174**(1):145-157

[35] Jailani AAK, Solanki V, Roy A, Sivasudha T, Mandal B. A CGMMV genome-replicon vector with partial sequences of coat protein gene efficiently expresses GFP in Nicotiana benthamiana. Virus Research. 2017;**233**:77-85

[36] Teoh PG, Ooi AS, AbuBakar S, Othman RY. Virus-specific read-through codon preference affects infectivity of chimeric cucumber green mottle mosaic viruses displaying a dengue virus epitope. Journal of Biomedicine & Biotechnology. 2009;**2009**:781712

[37] Ooi A, Tan S, Mohamed R, Rahman NA, Othman RY. The full-length clone of cucumber green mottle mosaic virus and its application as an expression system for hepatitis B surface antigen. Journal of Biotechnology. 2006;**121**(4):471-481

[38] Rhee SJ, Jang YJ, Lee GP. Identification of the subgenomic promoter of the coat protein gene of cucumber fruit mottle mosaic virus and development of a heterologous expression vector. Archives of Virology volume. 2016;**161**:1527-1538

[39] Chattopadhyay A, Abdul Kader Jailani A, Roy A, Mukherjee SK, Mandal B. Prediction of putative regulatory elements in the subgenomic promoters of cucumber green mottle mosaic virus and their interactions with the RNA dependent RNA polymerase domain. VirusDisease. 2020;**31**(4):503-516

[40] Shivprasad S, Pogue GP, Lewandowski DJ, Hidalgo J, Donson J, Grill LK, et al. Heterologous sequences greatly affect foreign gene expression in tobacco mosaic virus-based vectors. Virology. 1999;**255**(2):312-323

[41] Hongying, Zheng CX, Han K, Peng J, Lin L, Yuwen Lu LX, Xiaohua W, et al. Development of an agroinoculation system for full-length and GFP-tagged cDNA clones of cucumber green mottle mosaic virus. Archives of Virology. 2015;**160**(11):2867-2872

[42] Liu M, Liang Z, Aranda MA, Hong N, Liu L, Kang B, et al. A cucumber green mottle mosaic virus vector for virus-induced gene silencing in cucurbit plants. Plant Methods. 2020;**16**:9

[43] Xu A, Wei N, Hu H, Zhou S, Huang Y, Kong Q, et al. Thioredoxin h2 inhibits the MPKK5-MPK3 cascade to regulate the CBF–COR signaling pathway in Citrullus lanatus suffering chilling stress. Horticulture Research. 2022;**10**(2):256

[44] Guo H, Bi X, Wang Z, Jiang D, Cai M, An M, et al. Reactive oxygen species-related genes participate in resistance to cucumber green mottle mosaic virus infection regulated by boron in Nicotiana benthamiana and watermelon. Frontiers in Plant Science. 2022;**13**:1027404

[45] Wu J, Zhu M, Liu W, Jahan MS, Gu Q, Shu S, et al. CsPAO2 improves salt tolerance of cucumber through the interaction with CsPSA3 by affecting photosynthesis and polyamine conversion. International Journal of Molecular Sciences. 2022;**23**(20):12413

[46] Rhee SJ, Jang YJ, Park JY, Ryu J, Lee GP. Virus-induced gene silencing for in planta validation of gene function in cucurbits. Plant Physiology. 2022;**190**(4):2366-2379

[47] Li C, Yoshikawa N, Takahashi T, Ito T, Yoshida K, Koganezawa H. Nucleotide sequence and genome organization of apple latent spherical virus: A new virus classified into the family Comoviridae. The Journal of General Virology. 2000;**81**(Pt 2):541-547

[48] Yoshikawa N, Okada K, Asamuma K, Watanabe K, Igarasi A, Li C, et al. A movement protein and three capsid proteins are all necessary for the cell-to-cell movement of apple latent spherical cheravirus. Archives of Virology. 2006;**151**(5):837-848

[49] Yaegashi H, Yamatsuta T, Takahashi T, Li C, Isogai M, Kobori T, et al. Characterization of virus-induced gene silencing in tobacco plants infected with apple latent spherical virus. Archives of Virology. 2007;**152**(10):1839-1849

[50] Godge MR, Purkayastha A, Dasgupta I, Kumar PP. RETRACTED ARTICLE: Virus-induced gene silencing for functional analysis of selected genes. Plant Cell Reports. 2008;**27**(2):209-219

[51] Yamagishi N, Yoshikawa N. Efficient virus-induced gene silencing system in pumpkin (Cucurbita maxima) using apple latent spherical virus vector. Journal of Virological Methods. 2022;**301**:114456

[52] Fu DQ, Zhu BZ, Zhu HL, Jiang WB, Luo YB. Virus-induced gene silencing in tomato fruit. The Plant Journal. 2005;**43**(2):299-308

[53] Kannan M, Zainal Z, Ismail I, Baharum SN, Bunawan H. Application of reverse genetics in functional genomics of Potyvirus. Viruses. 2020;**12**(8):803

[54] Rodamilans B, Valli A, Mingot A, San León D, Baulcombe D, López-Moya JJ, et al. RNA polymerase slippage as a mechanism for the production of frameshift gene products in plant viruses of the potyviridae family. Journal of Virology. 2015;**89**(13):6965-6967

[55] Beauchemin C, Bougie V, Laliberté J-F. Simultaneous production of two foreign proteins from a potyvirus-based vector. Virus Research. 2005;**112**(1):1-8

[56] Kelloniemi J, Mäkinen K, Valkonen JPT. Three heterologous proteins simultaneously expressed from a chimeric potyvirus: Infectivity, stability and the correlation of genome and virion lengths. Virus Research. 2008;**135**(2):282-291

[57] Arazi T, Slutsky SG, Shiboleth YM, Wang Y, Rubinstein M, Barak S, et al. Engineering zucchini yellow mosaic potyvirus as a non-pathogenic vector for expression of heterologous proteins in cucurbits. Journal of Biotechnology. 2001;**87**(1):67-82

[58] Xu XJ, Zhu Q, Jiang SY, Yan ZY, Geng C, Tian YP, et al. Development and evaluation of stable sugarcane mosaic virus mild mutants for cross-protection against infection by severe strain. Frontiers in Plant Science. 2021;**12**:788963

[59] Gal-On A. A point mutation in the FRNK motif of the Potyvirus helper component-protease gene alters symptom expression in cucurbits and elicits protection against the severe homologous virus. Phytopathology. 2000;**90**(5):467-473

[60] Shiboleth YM, Arazi T, Wang Y, Gal-On A. A new approach for weed

control in a cucurbit field employing an attenuated potyvirus-vector for herbicide resistance. Journal of Biotechnology. 2001;**92**(1):37-46

[61] Shoresh M, Gal-On A, Leibman D, Chet I. Characterization of a mitogen-activated protein kinase gene from cucumber required for Trichoderma-conferred plant resistance. Plant Physiology. 2006;**142**(3):1169-1179

[62] Kang M, Seo JK, Choi H, Choi HS, Kim KH. Establishment of a simple and rapid gene delivery system for cucurbits by using engineered of zucchini yellow mosaic virus. Plant Pathology Journal. 2016;**32**(1):70-76

[63] Bertin S, Manglli A, McLeish M, Tomassoli L. Genetic variability of watermelon mosaic virus isolates infecting cucurbit crops in Italy. Archives of Virology. 2020;**165**(4):937-946

[64] Aragonés V, Pérez-de-Castro A, Cordero T, Cebolla-Cornejo J, López C, Picó B, et al. A watermelon mosaic virus clone tagged with the yellow visual maker phytoene synthase facilitates scoring infectivity in melon breeding programs. European Journal of Plant Pathology. 2019;**153**:1317-1323

[65] Houhou F, Aragonés V, Butković A, Sáez C, Picó B, López C, et al. A Virus Induced Gene-Silencing Vector Based on a Mild Isolate of Watermelon Mosaic Virus for Reverse Genetic Analyses in Cucurbits. 2021.

[66] Agaoua A, Rittener V, Troadec C, Desbiez C, Bendahmane A, Moquet F, et al. A single substitution in Vacuolar protein sorting 4 is responsible for resistance to watermelon mosaic virus in melon. Journal of Experimental Botany. 2022;**73**(12):4008-4021

[67] Wylie SJ, Adams M, Chalam C, Kreuze J, López-Moya JJ, Ohshima K, et al. Ictv Report Consortium. ICTV Virus Taxonomy Profile: Potyviridae. J Gen Virol. Mar 2017;**98**(3):352-354

[68] Ratcliff F, Martin-Hernandez AM, Baulcombe DC. Technical advance. Tobacco rattle virus as a vector for analysis of gene function by silencing. The Plant Journal. 2001;**25**(2):237-245

[69] MacFarlane SA. Molecular biology of the tobraviruses. The Journal of General Virology. 1999;**80**(Pt 11):2799-2807

[70] Zhang F, Chen C, Ge H, Liu J, Luo Y, Liu K, et al. Efficient soybean regeneration and agrobacterium-mediated transformation using a whole cotyledonary node as an explant. Biotechnology and Applied Biochemistry. 2014;**61**(5):620-625

[71] Bu R, Wang R, Wei Q, Hu H, Sun H, Song P, et al. Silencing of glycerol-3-phosphate acyltransferase 6 (GPAT6) gene using a newly established virus induced gene silencing (VIGS) system in cucumber alleviates autotoxicity mimicked by cinnamic acid (CA). Plant and Soil. 2019;**438**(1-2):329-346

[72] Liao JJ, Wang CH, Xing QJ, Li YP, Liu XF, Qi HY. Overexpression and VIGS system for functional gene validation in oriental melon (Cucumis melo var. makuwa Makino). Plant Cell, Tissue and Organ Culture (PCTOC). 2019;**137**(2):275-284

[73] Ma X-H, Wu T-L. Rapid and efficient regeneration in soybean [Glycine max (L.) Merrill] from whole cotyledonary node explants. Acta Physiologiae Plantarum. 2008;**30**:209-216

[74] Wei Y, Han X, Wang Z, Gu Q, Li H, Chen L, et al. Development of a GFP expression vector for cucurbit chlorotic yellows virus. Virology Journal. 2018;**15**(1):93

[75] Kubota K, Usugi T, Tsuda S. Production of antiserum and immunodetection of cucurbit chlorotic yellows virus, a novel whitefly-transmitted crinivirus. Journal of General Plant Pathology. 2011;**77**(2):116-120

[76] Harrison BD, Barker H, Bock KR, Guthrie EJ, Meredith G, Atkinson M. Plant viruses with circular single-stranded DNA. Nature. 1977;**270**(5639):760-762

[77] Kanakala S, Kuria P. Chickpea chlorotic dwarf virus: An emerging monopartite dicot infecting Mastrevirus. Viruses. 2018;**11**(1):5

[78] Zhang X, Mason H. Bean yellow dwarf virus replicons for high-level transgene expression in transgenic plants and cell cultures. Biotechnology and Bioengineering. 2006;**93**(2):271-279

[79] Huang Z, Phoolcharoen W, Lai H, Piensook K, Cardineau G, Zeitlin L, et al. High-level rapid production of full-size monoclonal antibodies in plants by a single-vector DNA replicon system. Biotechnology and Bioengineering. 2010;**106**(1):9-17

[80] Yamamoto T, Hoshikawa K, Ezura K, Okazawa R, Fujita S, Takaoka M, et al. Improvement of the transient expression system for production of recombinant proteins in plants. Scientific Reports. 2018;**8**(1):4755

[81] Wang Y, Wang J, Guo S, Tian S, Zhang J, Ren Y, et al. CRISPR/Cas9-mediated mutagenesis of ClBG1 decreased seed size and promoted seed germination in watermelon. Horticulture Research. 2021;**8**:70

[82] Igarashi A, Yamagata K, Sugai T, Takahashi Y, Sugawara E, Tamura A, et al. Apple latent spherical virus vectors for reliable and effective virus-induced gene silencing among a broad range of plants including tobacco, tomato, Arabidopsis thaliana, cucurbits, and legumes. Virology. 2009;**386**(2):407-416

[83] Butler NM, Baltes NJ, Voytas DF, Douches DS. Geminivirus-mediated genome editing in potato (Solanum tuberosum L.) using sequence-specific nucleases. Frontiers in Plant Science. 2016;**7**:1045

[84] Baltes NJ, Gil-Humanes J, Cermak T, Atkins PA, Voytas DF. DNA replicons for plant genome engineering. The Plant Cell. 2014;**26**(1):151-163

[85] Ali Z, Abul-faraj A, Piatek M, Mahfouz MM. Activity and specificity of TRV-mediated gene editing in plants. Plant Signaling & Behavior. 2015;**10**(10):e1044191

[86] Dahan-Meir T, Filler-Hayut S, Melamed-Bessudo C, Bocobza S, Czosnek H, Aharoni A, et al. Efficient in planta gene targeting in tomato using geminiviral replicons and the CRISPR/Cas9 system. The Plant Journal. 2018;**95**(1):5-16

[87] Avesani L, Marconi G, Morandini F, Albertini E, Bruschetta M, Bortesi L, et al. Stability of potato virus X expression vectors is related to insert size: Implications for replication models and risk assessment. Transgenic Research. 2007;**16**(5):587-597